T0140277

Springer Theses

Recognizing Outstanding Ph.D. Research

Aims and Scope

The series "Springer Theses" brings together a selection of the very best Ph.D. theses from around the world and across the physical sciences. Nominated and endorsed by two recognized specialists, each published volume has been selected for its scientific excellence and the high impact of its contents for the pertinent field of research. For greater accessibility to non-specialists, the published versions include an extended introduction, as well as a foreword by the student's supervisor explaining the special relevance of the work for the field. As a whole, the series will provide a valuable resource both for newcomers to the research fields described, and for other scientists seeking detailed background information on special questions. Finally, it provides an accredited documentation of the valuable contributions made by today's younger generation of scientists.

Theses are accepted into the series by invited nomination only and must fulfill all of the following criteria

- They must be written in good English.
- The topic should fall within the confines of Chemistry, Physics, Earth Sciences, Engineering and related interdisciplinary fields such as Materials, Nanoscience, Chemical Engineering, Complex Systems and Biophysics.
- The work reported in the thesis must represent a significant scientific advance.
- If the thesis includes previously published material, permission to reproduce this must be gained from the respective copyright holder.
- They must have been examined and passed during the 12 months prior to nomination.
- Each thesis should include a foreword by the supervisor outlining the significance of its content.
- The theses should have a clearly defined structure including an introduction accessible to scientists not expert in that particular field.

More information about this series at http://www.springer.com/series/8790

Sayan Biswas

Physics of Turbulent Jet Ignition

Mechanisms and Dynamics of Ultra-lean Combustion

 Springer

Sayan Biswas
School of Aeronautics and Astronautics
Purdue University
West Lafayette, IN, USA

ISSN 2190-5053   ISSN 2190-5061   (electronic)
Springer Theses
ISBN 978-3-030-09440-9   ISBN 978-3-319-76243-2   (eBook)
https://doi.org/10.1007/978-3-319-76243-2

Printed on acid-free paper

This Springer imprint is published by the registered company Springer International Publishing AG part of
Springer Nature.
The registered company address is: Gewerbestrasse 11, 6330 Cham, Switzerland

For Aditi Joshi
The joy of spending an entire lifetime
with such intelligence and beauty

Supervisor's Foreword

In the current effort of designing gas engines, which produce low exhaust emissions while maintaining power output and high efficiency, engine developers have turned increasingly to ultra-lean combustion technologies. For gas engine manufacturers and automotive industry, natural gas (liquefied and compressed) is a promising fuel due to its abundant supply, cheap cost, and adaptability as gas engine fuels. Thus, use of natural gas and its derivatives are rising. Currently, more than one billion motorized vehicles are running on gaseous fuels. Lean combustion strategies have the potentials to overcome the stringent emission regulations for oxides of nitrogen (NOx), unburned hydrocarbon, and particulate emission while increasing the fuel economy. Reliable operation of gas engines at the lean condition, however, gives rise to some serious challenges. Poor ignition or failure to ignition can lead to misfires, which can result in undesirable effects such as cycle-to-cycle variability, rough operation, and reduction in efficiency and increased unburned hydrocarbon emissions. Also, at lean limit flame becomes more susceptible to thermal-acoustic instabilities.

Pre-chamber turbulent jet ignition is an advanced ignition method and has received increasing interest for its application in medium/heavy-duty on-road natural gas engines. A small quantity of stoichiometric fuel/air is burned in a separate small-volume combustion chamber called the pre-chamber. The combustion products are then discharged into the main chamber filled with ultra-lean premixed fuel/air through a small diameter orifice in the form of a hot turbulent jet. Compared to a conventional spark plug, the hot jet has a much larger surface area leading to multiple ignition sites on its surface, which can enhance the probability of successful ignition and cause faster flame propagation and heat release. Thus, hot jet ignition has the potential to enable the combustion system to operate near the fuel's lean flammability limit, leading to ultra-low emissions.

However, the physics behind ultra-lean jet ignition is rather complicated. Several interrelated chemical and physical processes are involved. The complex coupling between turbulence and chemistry needs to be understood. Ignition mechanism heavily depends on the mixing characteristics and the composition of the hot jet.

The competition between thermal and mass diffusivity between the hot exhaust gases and the cold fresh lean mixtures and presence of active radicals such as H, O, HO_2, and OH all affect the ignition process. Biswas's thesis contributes to improving our knowledge and understanding of turbulent hot jet ignition phenomenon, from a fundamental point of view.

Biswas's thesis integrates experiments and simulations to understand the fundamental mechanisms and combustion dynamics of turbulent jet ignition systems. For example, to remove any geometric and thermodynamic parametric effects, a nondimensional Damköhler number is introduced to separate different ignition regimes. Additionally, the thesis discovers that the lean limit of fuel/air mixtures can be extended using a hot supersonic jet. The effect of multiple turbulent jets, which is relevant to practical engines, is thoroughly studied. Ignition due to jet impingement is addressed in detail. The characteristics and behavior of the combustion instability at the ultra-lean limit are carefully investigated. Furthermore, Biswas's thesis includes a novel Schlieren-based velocity measurement technique for high-speed turbulent flows. Overall, his thesis provides important insights and guidelines for future advancement and optimization of the pre-chamber turbulent jet ignition systems for ultra-lean natural gas engines.

West Lafayette, IN, USA   Li Qiao
January 2018

Preface

This book talks about the underlying physics of turbulent jet ignition, with a particular emphasis on the fundamental mechanisms and dynamics that occur at the intersection of the combustion, fluid mechanics, turbulence, and chemical kinetics disciplines. The material covered in this book is from my doctoral study at Purdue University. My doctoral research was focused on developing novel lean combustion strategies for sustainable energy development, improved fuel economy, and lesser pollutant emission. To achieve lean combustion a hot turbulent jet was used to ignite ultra-lean methane/air and hydrogen/air mixtures. Ignition of ultra-lean mixtures by a hot jet can be utilized in various applications ranging from pulse detonation engines, wave rotor combustor explosions, to supersonic combustors and natural gas engines. Compared to a conventional spark plug, the hot jet has a much larger surface area. This can lead to multiple ignition sites on its surface which can enhance the probability of successful ignition and cause faster flame propagation and heat release. The physics behind ultra-lean jet ignition is extremely complicated due to the complex coupling between turbulence and chemistry. My research aimed to unveil the complex physics behind turbulent jet ignition.

This book is divided into nine chapters. Chapter 1 introduces the concept of turbulent jet ignition, common challenges for lean operation, and discusses the existing literatures on turbulent jet ignition. Chapter 2 discusses different ignition mechanisms of methane/air and hydrogen/air mixtures by a hot turbulent jet. A nondimensional Damköhler number was introduced to remove parametric dependency and to separate different ignition regimes. Chapter 3 proposed a novel, two-camera approach for Schlieren-based image correlation velocimetry technique, named as Schlieren image velocimetry (SIV), to measure high-speed flows. Experimental measurement of highly transient high-speed turbulent flows is extremely challenging due to lack of accurate flow tracers. SIV eliminates the requirement of any flow tracer. The idea behind developing SIV was to resolve the velocity field of turbulent reacting jets that held the key to understand turbulent jet ignition. Chapter 4 discusses ignition of the ultra-lean mixture by a supersonic jet. Most studies conducted so far used a subsonic or near-sonic jet for ignition. Supersonic jets may have advantages in terms of more reliable ignition and faster burning and thus could potentially improve combustion efficiency. This led to the idea of using a supersonic hot jet to ignite ultra-lean fuel/air mixture. Chapter 5 addresses thermoacoustic instability. Thermoacoustic instability becomes severe at the ultra-lean limit. Controlling instability, in active or passive manner, requires adequate knowledge about different types of instability modes, perturbation energy, and frequencies associated with the instability. Chapter 6 discusses ignition by multiple hot turbulent jets. Ignition by multiple jets was motivated by the fact that most gas engine companies use multiple nozzle pre-chamber. Chapter 7 discusses ignition by impinging hot turbulent jet. The effect of impingement, confinement, and wall effect becomes predominant inside the small squeeze volume of the engine combustor. The hot jet issued from the pre-chamber may impinge onto the surface of the piston head or the wall of the main engine during the cycle. Chapter 8 addresses the flame propagation through straight and converging-diverging microchannels. The connecting nozzles between the pre-chamber and the main chamber play a key role in determining the ignition mechanism. Thus, to understand different ignition mechanisms, it is necessary to examine the flame propagation through narrow channels. This study helped to understand the details of ignition processes by subsonic and supersonic hot jet. Flame-wall interaction and molecular diffusion increases at microscale. Thus, the focus was to investigate flame extinction behavior at preeminent heat loss at the microscale. The last Chap. 9 summarizes current studies on turbulent jet ignition and talks about the future research directions.

My quest to explore the physics of turbulent jet ignition has not ended with my doctoral studies. Presently, I am investigating different plasma igniters and turbulent jet igniters at Engine Combustion Division of Sandia National Laboratories in Livermore, California. My goal is to understand the turbulent jet ignition processes inside an optically accessible light and medium duty automotive engine. I plan to develop prototype pre-chambers and jet igniters for leading automotive manufacturers. I believe turbulent jet ignition is the potential solution for enhanced fuel economy and emission reduction for next generation of gas engines.

For my doctoral research, I want to thank my advisor Dr. Li Qiao and the committee members Dr. Robert Lucht, Dr. Jay Gore, Dr. Haifeng Wang for their advice, guidance, and support. Sincere and heartfelt thanks to my labmates, summer undergraduate students, technicians, and safety manager. The financial support from Caterpillar Inc. and Purdue's School of Aeronautics and Astronautics is gratefully acknowledged. My deepest gratitude for my father Late Sanat Biswas and mother Chaina Biswas. I reserve my profound appreciation and gratitude for Aditi Joshi, whose sacrifice and endless love has been instrumental during my doctoral studies.

Livermore, CA, USA   Sayan Biswas
January 2018

Contents

Symbols, Subscripts, and Abbreviations

List of Symbols

English symbols

A	Area
$\bar{A}$	Arrhenius constant
C	Concentration
d, D	Diameter
Da	Damköhler number
$E, \mathbb{E}$	Energy
f	Frequency
$\mathbb{F}$	Fourier operator
g	Gravitational acceleration
h	Heat transfer coefficient
H	Impingement distance
$\mathbb{H}$	Enthalpy
i	$\sqrt{-1}$
I	Turbulent intensity
$\mathbb{I}$	Radiation intensity
k	Thermal conductivity
$\Bbbk$	Wave number
Ka	Karlovitz number
l	Integral length scale
L	Characteristic length
Le	Lewis number
m	Mass flow
M	Mach number
MW	Molecular weight

$\mathcal{M}$	Markstein number
N	Sample size
Nu	Nusselt number
p	Pressure
P	Probability
$\mathbb{P}$	Power
$\mathbb{Q}$	Heat release
r	Radial coordinate
R	Gas constant
Re	Reynolds number
s	Flame speed
$\mathbb{S}$	Source term
t	Time
T	Temperature
u, v, w	Components of velocity in Cartesian coordinate
U	Internal energy
$\mathbb{V}$	Volume
$\mathbb{W}$	Weight
x, y	Cartesian coordinate
X	Mole fraction
Y	Mass fraction
Z	Mixture fraction

Greek symbols

α	Angle
β	d/D
γ	Ratio of specific heats
δ	Small quantity
ε	Stress
ϵ	Angle of refraction
θ	Impinging angle
ϑ	Kinematic viscosity
μ	Dynamic viscosity
ρ	Density
κ	Strain rate
σ	Area ratio
τ	Ignition delay
ϕ	Equivalence ratio
ω	Complex potential
$\eta - \xi$	Local coordinate
Δ	Differential change
λ	Spectral
$\sum$	Flame surface density
ζ	$Y_i/Y_i \mid _{unburned}$

List of Subscripts

avg	Average
b	Blackbody, burned
crit	Critical
d	Diameter
f	Final, flame, fluid
i	Initial, i^{th} species
L	Laminar
main	Main chamber
mean	Mean
mix	Mixture
p	Particle
pre	Pre-chamber
tran	Transition
ub	Unburned
0	Centerline
1, 2	Pre-chamber and main chamber

List of Abbreviations

BOS	Background Oriented Schlieren
CCD	Charge Coupled Device
CD	Convergent Divergent
CWO	Continuous Window Offset
DMD	Dynamic Mode Decomposition
DNS	Direct Numerical Simulation
FFT	Fast Fourier Transform
HWP	Hot Wire Pyrometry
IR	Infrared
LDV	Laser Doppler Velocimetry
LES	Large Eddy Simulation
LIF	Laser Induced Fluorescence
ND	Neutral Density
PAH	Polycyclic Aromatic Hydrocarbon
PCE	Peak to Correlation Energy
PDF	Probability Distribution Function
PIV	Particle Image Velocimetry
POD	Proper Orthogonal Decomposition
PSD	Power Spectral Density
RANS	Reynolds Averaged Navier Stokes
RMS	Root Mean Square
ROI	Region of Interest
RPC	Robust Phase Correlation

PPR   Primary Peak Ratio
SIMPLE   Semi-Implicit Method for Pressure Linked Equations
SIV   Schlieren Image Velocimetry
SNR   Signal to Noise Ratio
UOD   Universal Outlier Detector

Chapter 1
Introduction

Contents

1.1 Greenhouse Emission

Greenhouse gases trap heat and make the planet warmer. According to the Environmental Protection Agency (EPA), the four major greenhouse gases (GHG) are CO_2, methane, oxides of nitrogen (NOx), and fluorinated gases such as hydrofluorocarbon, perfluorocarbon, sulfur hexafluoride, nitrogen trifluoride, etc. [1]. Figure 1.1 shows US greenhouse gas emissions in the years 1990–2015. Human activities are responsible for almost all the increase in greenhouse gases in the atmosphere over the last 150 years [2, 3]. Figure 1.2 shows the greenhouse gas emissions in the USA in 2015. One of the largest contributors toward greenhouse emission is carbon dioxide. Major production of CO_2 is from burning fossil fuels such as coal, natural gas, and oil. Methane is emitted during the production and transport of coal, natural gas, and oil [4].

The largest source of greenhouse gas emissions from human activities in the USA is from burning fossil fuels for electricity, heat, and transportation [6]. EPA tracks total US greenhouse emission from various economic sectors such as energy, transportation, residential, agriculture, etc.

Figure 1.3 shows the greenhouse gas emission from different economic sectors in 2015. Transportation, industry, and electricity are the three major sectors

© Springer International Publishing AG, part of Springer Nature 2018
S. Biswas, *Physics of Turbulent Jet Ignition*, Springer Theses,
https://doi.org/10.1007/978-3-319-76243-2_1

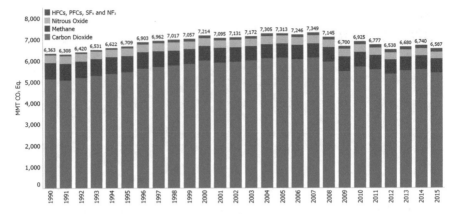

Fig. 1.1 Gross US greenhouse gas emissions in million metric tons (MMT) of CO_2 equivalent from 1990 to 2015 [5]

Fig. 1.2 US greenhouse gas emissions in 2015 [5]

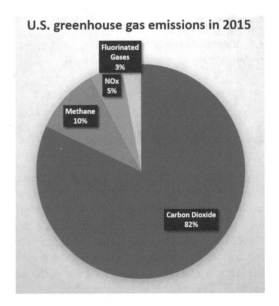

contributing to the US greenhouse gas emissions. Total emissions in 2015 was 6587 million metric tons (MMT) of CO_2 equivalent [5].

The main driver of emissions in the electricity, industry, and transportation sector is from fossil fuel combustion [7]. Fossil fuels are petroleum, coal, natural gas, and geothermal sources, formed from the remains of dead plants and animals. Figure 1.4 describes the total emissions from fossil fuel combustion, separated by end-use sector, including CH_4 and oxides of nitrogen in addition to CO_2. In 2015, a staggering 82% of total US greenhouse gas emissions were from fossil fuel combustion [5, 8]. GHG is an eminent threat to the public health and environment [9].

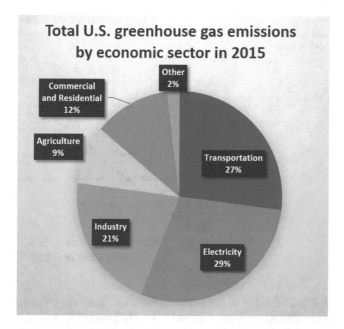

Fig. 1.3 Total US greenhouse gas emissions by economic sector in 2015 [5]

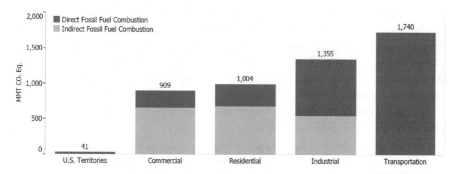

Fig. 1.4 End-use sector emissions of CO_2 from fossil fuel combustion in 2015 (MMT CO_2 equivalent) [5]

1.2 Emission Regulation and Penalties

The US Environmental Protection Agency (EPA) is adopting nonconformance penalties (NCPs) for heavy heavy-duty diesel engines that can be used by manufacturers of heavy-duty diesel engines unable to meet the current oxides of nitrogen (NOx) emission standard [10]. These penalties, which are assessed on a per engine basis, allow a manufacturer to produce and sell nonconforming engines upon payment of penalties. The actual penalties reflect how close the engines are to

Fig. 1.5 Heavy-duty engine nonconformance penalties per engine basis [11]

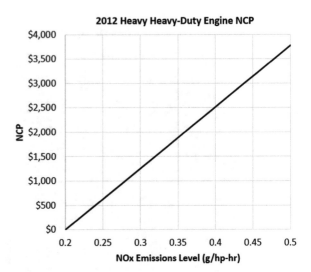

meeting the standard – the cleaner the engines are, the lower the penalties will be. EPA has established an upper limit of NOx emission to 0.5 g/hp-h for heavy-duty engines by the end of 2012. Figure 1.5 shows the nonconformance penalties for heavy-duty engines based on NOx emission level. An engine manufacturer must pay $3775/engine at maximum NOx emission limit, 0.5 gm/hp-h [11, 12]. Further stringent rules and regulations on emission will be adopted in future years.

US emission regulations from the transportation sector, such as those established by the state of California, have become more aggressive regulations, especially on mobile source emission. A central concept of the emission standard setting process for mobile source emissions by EPA and California Air Resources Board (CARB) is termed as "technology forcing" [13]. It has been applied to a wide range of sources, including light-duty vehicles, on-road diesel engines, and nonroad engines, for the control of CO_2, CO, UHCs, and NOx.

> ... "Technology forcing" refers to the establishment by a regulatory agency of a requirement to achieve an emissions limit, within a specified time frame, that can be reached through use of unspecified technology or technologies that have not yet been developed for widespread commercial applications and have been shown to be feasible on an experimental or pilot-demonstration basis [13].

A potential solution for engine manufacturers to these stringent emission regulations for greenhouse gases is to implement lean combustion strategies [14]. Lean operation reduces the peak combustion temperature, thereby minimizing thermal NOx formation. The amount of CO_2 produces from fossil fuels is reduced as well as UHC emissions. Soot particulate and water vapor emission are extremely small during the lean-burn. Thus, switching to lean combustion can alleviate the problem of greenhouse emission.

1.3   Ultra-lean Combustion Strategies

Today lean combustion strategies are used and employed in nearly all combustion technology sectors, including gas turbines, internal combustion engines, industrial furnaces, stationary power sector, etc. [14, 15]. This is because combustion processes operating under fuel-lean conditions can have very low emissions and very high efficiency. Moreover, greenhouse gases and particulate emissions are reduced since flame temperatures are typically low in lean-burn conditions, reducing thermal NOx formation. Also, lean hydrocarbon combustion reduces unburned hydrocarbon and CO emissions. Unfortunately, operating at lean condition is challenging. Implementing the lean combustion strategies in practical combustion systems is limited by low reaction rates, ignition probability, cycle-to-cycle variability, limited knowledge in low-temperature chemistry, extinction, combustion instabilities, mild heat release, sensitivity to mixing, etc. [16, 17].

Another great challenge in lean combustion is the initiation of ignition. Ignition becomes increasingly difficult as the fuel/air mixture becomes leaner [18, 19]. Figure 1.6 shows the ignitability curve of methane/air at atmospheric pressure and temperature.

At lean conditions, the ignition energy required to initiate the ignition increases sharply. Reliable ignition becomes difficult at fuel-lean conditions using a conventional spark plug. Since conventional spark ignition system produces a high-temperature zone confined at a certain location of the combustor, as the combustor mixture goes leaner, the ignition probability drops. It can adversely affect the combustion performance. Poor ignition or failure to ignite the lean mixture can lead to misfires, which can lead to undesirable effects such as cycle-to-cycle variability, rough operation, and reduction in efficiency and increased unburned hydrocarbon emissions [14].

Fig. 1.6 Ignitability curve for methane/air mixtures at atmospheric conditions [19]

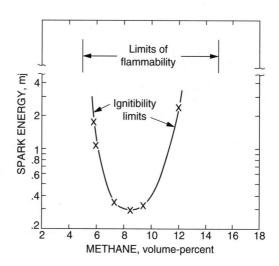

Few ways to perform a reliable ultra-lean ignition are hot turbulent jet ignition [20], plasma ignition [21, 22], laser ignition [23, 24], etc. All these ignition techniques have several potential advantages over conventional spark ignition systems. However, the hot turbulent jet ignition is the most attractive among these ultra-lean ignition strategies due to its conceptual simplicity and minimum design change required to implement in current engines.

1.4 Ignition as Limit Phenomena

From a theoretical perspective, ignition can be achieved in one of two ways [25]. The first way to initiate ignition is by supplying a momentary or continuous amount of heat such as by an electric spark to the combustible mixture. The heated fuel/air mixture responds in Arrhenius fashion and produces more heat. However, with an increase in temperature, the mixture tends to lose more heat to the cooler part of the mixture and the combustor wall. If the rate of heat generation is greater than the cooling losses, then the thermal runway eventually leads to ignition. Semenov [26] discussed the behavior of such ignition by comparing the relative magnitudes of the heat generation term to the nonlinear heat loss terms as shown in Fig. 1.7.

The second way to initiate ignition is by providing a sufficient number of chain-branching radicals. As the active radicals undergo branching, if the initiation and chain-branching reactions dominate over the chain-termination reactions, it leads to a sustainable thermal runway, and finally, ignition occurs. This type of ignition occurs when a jet full of active radicals ignites a combustible mixture. Our current work focuses on hot turbulent jet ignition. Ignition by a hot turbulent jet has several advantages over conventional spark ignition and will be discussed in detail in the following sections.

Fig. 1.7 The Semenov criterion is showing how the ignitability of a mixture depends on the competition between the nonlinear heat generation rate and linear heat loss rate [27]

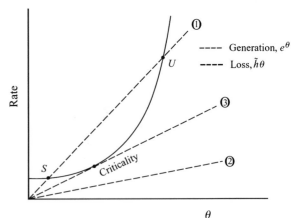

1.5   Natural Gas and Hydrogen

Natural gas [28, 29] and hydrogen [30, 31] both have the potential to be used in ultra-lean combustion technologies for major reasons. Natural gas, composed chiefly of methane, is considered as the cleanest fossil fuel. Natural gas burns cleaner than other fossil fuels, producing half the carbon dioxide as coal and about a third less than oil, which is the primary greenhouse gas. It also emits fewer amounts of toxic chemicals like nitrogen oxides, sulfur dioxide, and particulate matter. In the US market, natural gas is an abundant resource. Another major benefit of natural gas is that it can be stored efficiently and safely.

Hydrogen is an environmentally friendly alternative to fossil fuels. The most important feature of hydrogen energy is that it is a zero-carbon-footprint fuel. When hydrogen is burned, it leaves no trace or residue that would affect the public health and environment. Hydrogen combustion emits only water vapor. Hydrogen is renewable, readily available, and most importantly far more efficient than other energy sources. Thus, natural gas and hydrogen are two potential fuels of our future energy system [31, 32].

1.6   Background on Hot Turbulent Jet Ignition

In the current effort of designing gas engines which produce lower exhaust emissions while maintaining power output and high efficiency, engine developers have turned increasingly to the use of advanced combustion technologies and alternative fuels. Gaseous fuels, in general, and natural gas, in particular, are promising alternative fuels due to their abundant natural supply, economical cost, and adaptability as engine fuels. It is reported (January 2016) that in the transportation sector, worldwide, more than 1.1 vehicle vehicles (passenger cars, buses, and trucks) are running every day on gaseous fuels [33]. With the incentives of high fuel prices, stringent emission regulations for oxides of nitrogen (NOx), and focus on guaranteed fuel supplies, considerable research has been conducted in utilizing lean combustion potentials and overcoming its associated problems. Over the years, lean combustion engines have shown lower emissions and fuel consumption although unburned hydrocarbon emissions sometimes increased [15, 34]. The ultra-lean operation, however, gives rise to some serious challenges. As the fuel/air mixture becomes leaner, ignition becomes more difficult. Poor ignition or failure to ignite the lean mixture can lead to misfires, which can result in undesirable effects such as cycle-to-cycle variability, rough operation, and reduction in efficiency and increased unburned hydrocarbon emissions [35]. To achieve the demand of smooth performance and good acceleration, the abovementioned problems have made it difficult and challenging to utilize the benefits of lean mixtures unless changeover in ultra-lean combustion technologies is made [36, 37].

An approach that can potentially solve these problems is to use a hot turbulent jet to ignite the ultra-lean mixture. The ignition of fuel/air mixtures by a hot jet is a process utilized in various applications ranging from pulse detonation engines, wave rotor combustor explosions, to supersonic combustors and natural gas engines [38–40]. The hot turbulent jet is generated by burning a small quantity of stoichiometric or near-stoichiometric fuel/air mixture in a separate small volume called the pre-chamber. The higher pressure resulting from pre-chamber combustion pushes the combustion products into the main chamber in the form of a hot turbulent jet, which then ignites the ultra-lean premixed fuel/air in the main chamber. Compared to a conventional spark plug, the hot jet has a much larger surface area leading to multiple ignition sites on its surface which can enhance the probability of successful ignition and cause faster flame propagation and heat release. Pre-chamber-generated turbulent jet ignition has not received as much attention as the other attractive alternative to spark ignition, diesel pilot injection. This can partially be explained by the complex nature of the ignition resulting from a pre-chamber ignition system. Thus, a detailed investigation to understand complex physical processes behind turbulent jet ignition was necessary.

1.7   Literature Review

Some of the earliest known works on hot jet ignition are done by Wolfhard [41], Gussak [42–44], Murase [37, 45], and Oppenheim [46, 47]. Wolfhard [41] used a continuously injected hot gas jet heated in a ceramic furnace to ignite cold explosive mixtures. In 1966, Russian scientist Goossak Lev Abramovich (L. A. Gussak) proposed the use of a very rich mixture ($\phi = 1.4 - 2.5$) in a small separate chamber called pre-chamber to produce a low-temperature reacting jet filled with chemically active combustion products containing species like CO, H_2, aldehyde, and peroxide. This concept was called in Russian "Lavinia Aktyvatsia Gorenia" and hence is generally referred to as the LAG ignition process. During the late 1980s, Oppenheim offered a pre-chamber combustion technique similar to LAG named Pulsed Jet Combustion (PJC). Using PJC the lean limit marginally extended to $\phi = 0.54$.

Ghoniem and Chen [48] was among the first to investigate the fundamental mechanisms of hot turbulent jet ignition. It was found that the formation of the large-scale eddy structure of the turbulent jet was first triggered by vortex pairs. These eddy structures dominated the external features of the jet. The flame shape, speed, and propagation processes had a strong dependence on the number and the locations of the ignition sources deposited onto the turbulent jet. Pitt [49] showed the jet ignition system exhibits shorter delay times and increased burn rates when compared to a conventional spark system. Yamaguchi [50] investigated the effect of orifice diameter in a divided chamber bomb. Their results showed that a smaller orifice diameter resulted in "well-dispersed burning," the larger orifice diameter resulted in "flame kernel torch ignition," and the largest orifice diameter enabled laminar flame to pass through the orifice. Wallesten [51] explored the effect of spark position in the pre-chamber in relation to an orifice on the combustion efficiency of the main chamber. They found the spark position farther away from the orifice showed higher combustion rate than the positions close to the orifice. Elhsnawi [52] explored ignition of near-stoichiometric H_2/O_2 mixture by a hot inert gas jet (argon and nitrogen) in a detonation tube with an orifice size ranging from 8 mm to 11.2 mm. It was observed that the main chamber ignition initiates from jet side surfaces in the turbulent mixing zone. Sadanandan [53, 54] studied ignition of H_2/air mixtures by a hot jet using a high-speed laser schlieren and OH planar laser-induced fluorescence (PLIF) techniques. They observed no appreciable amount of OH radicals at the orifice exit and speculated that possible heat loss through the orifice walls had a strong influence. Additionally, they found that ignition in the main chamber occurs near the jet tip and not at the lateral sides of the jet.

Studies recently conducted by Toulson and Gholamisheeri [20, 55–58] and Attard [59–61] were focused on the effect of different pre-chamber fuels (H_2, C_3H_8, natural gas, CO) on combustion stability, extension of lean limit, and emission control in an optically accessible engine. The lean operation enabled by the turbulent jet ignition system resulted in near elimination of in-cylinder NOx emissions; significant improvements in efficiency and fuel economy were observed. Perera [62] investigated the ignitability and ignition delay time for ethylene/air ignited by a hot jet and found a lean equivalence ratio limit of 0.4 for the main chamber mixture. Carpio [63] numerically studied the critical radius of an axisymmetric jet comprising combustion products for ignition of H_2/air using detailed chemistry. For a given equivalence ratio, the critical radius is found to increase with increasing injection velocities. On the other hand, for a given injection velocity, the smallest critical radius is found at stoichiometric conditions. Karimi [64] numerically studied ignition of ethylene/air and CH_4/air in a long constant volume combustor using a traversing hot jet. They found that a higher traverse rate of the hot jet increases delay time and lowers the entrainment rate and in turn jet-vortex interaction. Shah [65, 66] studied the effect of pre-chamber volume and orifice diameter for heavy-duty natural gas engines. They found that increasing the pre-chamber to main chamber volume ratio does not significantly increase the effectiveness of the pre-chamber as an ignition device. Orifice diameter was not found to have an effect on the lean limit except for the largest pre-chamber volume cases.

1.8 Research Motivation and Objectives

The underlying physics of hot turbulent jet ignition, however, is a complicated phenomenon. There are very few computational and experimental studies on the fundamental mechanisms involving the complex coupling between turbulent mixing and chemical reactions. Several interrelated chemical and physical processes are involved. For example, the jet containing hot combustion products penetrate into the lean mixture, providing a high-temperature environment for mixing and ignition. It may also contain active radicals such as H, O, and OH, which initiate chain-branching reactions [44]. We can expect that the radicals which are important for ignition chemistry, the mixing process between the hot exhaust and the fresh lean mixtures, turbulence, and strain rate all affect the ignition process.

Many of the studies mentioned above, especially those conducted using an optically accessible engine or a real engine, show the great potential for hot jet ignition for extra-lean combustion. Our fundamental understanding, however, is far from complete. For example, there are contradictions in the literature on the ignition locations. Sadanandan [53, 54] observed ignition occurring near the tip of the jet, whereas Elhsnawi [52] observed ignition occurring on the lateral sides of the jet. Furthermore, depending on pre-chamber geometry and operating conditions, main chamber ignition can be caused by either a flame jet containing active radicals or a hot jet containing combustion products without any radicals. For example, Toulson [56–58] found that a jet torch is responsible for ignition although it was not clear whether the torch contained active radicals.

Most studies conducted so far used a subsonic or near-sonic jet for ignition. Supersonic jets may have advantages in terms of more reliable ignition and faster burning and thus could potentially improve combustion efficiency. This led to the idea of using a supersonic hot jet to ignite ultra-lean fuel/air mixture. Also, we wanted to investigate if ignition delay gets shorter due to the high velocity of supersonic jets.

Thermoacoustic instability becomes severe at the ultra-lean limit. Current research objective also includes investigation of thermoacoustic instability in the hot jet ignition of ultra-lean premixed H_2/air. Our interest was to study the unstable modes arising from thermoacoustic instability and their behavior, growth and decay, influence of equivalence ratio on instability growth, and flame dynamics at instability.

Complete understanding and knowledge of the physical processes behind jet ignition phenomena are not available at this time. Moreover, combustion instability triggers at the lean limit. Definite values for the safe lean operating limit, which represents the upper bound of the operational range of lean-burn without significant combustion instability, are not available in the literature for pre-chamber-generated jet ignition. Therefore, the main objective of this study is to investigate in detail the fundamental mechanisms behind turbulent jet ignition and to examine combustion dynamics near the ultra-lean limit.

References

1. United States: Environmental Protection Agency. Office of Policy, P., and evaluation. In: Inventory of U.S. Greenhouse Gas Emissions and Sinks: 1990–2014. U.S. Environmental Protection Agency, Washington, DC (2016)
2. Auxt, J.A., Curtis, W.M.: Global Warming : and the Creator's Plan, p. 169. Master Books, Green Forest (2009)

3. Officer, C.B., Page, J., Officer, C.B. (eds.): When the Planet Rages : Natural Disasters, Global Warming, and the Future of the Earth. Rev. and updated ed, vol. xvii, p. 227. Oxford University Press, Oxford, UK/New York (2009)

4. Intergovernmental Panel on Climate Change: Climate Change : the IPCC Response Strategies, vol. lxii, p. 272. Island Press, Washington, DC (1991)

5. United States: Environmental Protection Agency. In: Office of Policy, P., and Evaluation., Inventory of U.S. Greenhouse Gas Emissions and Sinks: 1990–2015. U.S. Environmental Protection Agency, Washington, DC (2017)

6. Easton, T.A.: Environmental studies. In: Classic Edition Sources, vol. xxi, 3rd edn, p. 228. McGraw-Hill Higher Education, New York (2009)

7. Parry, M., Canziani, O., Palutikof, J., Linden, P., Hanson, C.: Climate Change. In: Impacts, Adaptation and Vulnerability, pp. 10013–12473. Cambridge University Press, New York (2007)

8. GAO: Legislative Branch: Energy Audits are Key to Strategy for Reducing Greenhouse Gas Emissions. United States Government Accountability Office. GAO-07-516 (2007)

9. Woodcock, J., Phil, E., Tonne, C., Armstrong, B.G., Ashiru, O., Banister, D., Beevers, S., Chalabi, Z., Chowdhury, Z., Cohen, A., Franco, O.H., Haines, A., Hickman, R., Lindsay, G., Mittal, I., Mohan, D., Tiwari, G., Woodward, A., Roberts, I.: Public health benefits of strategies to reduce greenhouse-gas emissions: urban land transport. Lancet. **374**(9705), 1930–1943 (2009)

10. United States, E.P.A: Office of Transportation and Air Quality, Non-conformance Penalties for Heavy-Duty Diesel Engines Subject to the 2010 NOx Emission Standard. U.S. Environmental Protection Agency, Office of Transportation and Air Quality, Washington, DC (2012)

11. Caterpillar: Cat estimates \$40 million expense for EPA non-conformance penalties (NCPs) for post-October, 2002 diesel engines. Diesel Fuel News. **7**(5), 7–10 (2003)

12. Peckham, J.: U.S. EPA doubles 'non-conformance penalties' for navistar 2012 diesel trucks. Diesel Fuel News. **16**(34), 4–5 (2012)

13. Standards., N.R.C.U.S.B.o.E.S.a.T.N.R.C.U.S.C.o.S.P.i.S.M.S.E: State and Federal Standards for Mobile Source Emissions. National Academies Press, Washington, DC (2006)

14. Dunn-Rankin, D.: Lean Combustion : Technology and Control, vol. xi, p. 261. Academic Press, Amsterdam/Boston (2008.) 8 p. of plates

15. Kim, K., et al.: Evaluation of injection and ignition schemes for the ultra-lean combustion direct-injection LPG engine to control particulate emissions. Appl. Energy. **194**, 123–135 (2017)

16. Lieuwen, T.C., Yang, V.: Combustion instabilities in gas turbine engines : operational experience, fundamental mechanisms and modeling. In: Progress in Astronautics and Aeronautics, vol. xiv, p. 657. American Institute of Aeronautics and Astronautics, Reston (2005)

17. Society of Automotive Engineers: Homogeneous Charge Compression Ignition (HCCI) Combustion 2004, p. 280. Society of Automotive Engineers, Warrendale (2004)

18. Zabetakis, M.G., Instrument Society of America: Installation, Operation, and Maintenance of Combustible Gas Detection Instruments : Recommended Practice, p. 153. Instrument Society of America, Research Triangle Park (1987)

19. Zabetakis, M.G.: Flammability Characteristics of Combustible Gases and Vapors, vol. vii, p. 121. U.S. Dept. of the Interior, Bureau of Mines; for sale by the Superintendent of Documents, U.S. Govt. Print. Off, Washington (1965)

20. Gholamisheeri, M., Wichman, I.S., Toulson, E.: A study of the turbulent jet flow field in a methane fueled turbulent jet ignition (TJI) system. Combust. Flame. **183**, 194–206 (2017)

21. Wang, Z., et al.: Experimental study of microwave resonance plasma ignition of methane–air mixture in a constant volume cylinder. Combust. Flame. **162**(6), 2561–2568 (2015)

22. Guan, Y., Zhao, G., Xiao, X.: Design and experiments of plasma jet igniter for aeroengine. Propuls. Pow. Res. **2**(3), 188–193 (2013)

23. An, B., et al.: Experimental investigation on the impacts of ignition energy and position on ignition processes in supersonic flows by laser induced plasma. Acta Astronaut. **137**, 444–449 (2017)
24. Li, X., et al.: Experimental investigation on laser-induced plasma ignition of hydrocarbon fuel in scramjet engine at takeover flight conditions. Acta Astronaut. **138**, 79–84 (2017)
25. Glassman, I., Yetter, R.A.: Combustion, vol. xx, 4th edn, p. 773. Academic Press, Amsterdam/ Boston (2008)
26. Semenov, N.N.: Some Problems in Chemical Kinetics and Reactivity, 2nd edn. Princeton University Press, Princeton (1958)
27. Law, C.K.: Combustion Physics, vol. xviii, p. 722. Cambridge University Press, Cambridge, UK/New York (2006)
28. Healey, S., Jaccard, M.: Abundant low-cost natural gas and deep GHG emissions reductions for the United States. Energy Policy. **98**, 241–253 (2016)
29. Arezki, R., Fetzer, T., Pisch, F.: On the comparative advantage of U.S. manufacturing: evidence from the shale gas revolution. J. Int. Econ. **107**, 34–59 (2017)
30. Gurz, M., et al.: The meeting of hydrogen and automotive: a review. Int. J. Hydrog. Energy. **42** (36), 23334–23346 (2017)
31. Reddi, K., et al.: Building a Hydrogen Infrastructure in the United States, pp. 293–319 (2016) https://www.sciencedirect.com/science/article/pii/B9781782423645000130; https://doi.org/10. 1016/B978-1-78242-364-5.00013-0
32. Talus, K.: United States natural gas markets, contracts and risks: what lessons for the European Union and Asia-Pacific natural gas markets? Energy Policy. **74**, 28–34 (2014) https://www. sciencedirect.com/science/article/pii/S0301421514004510; https://doi.org/10.1016/j.enpol. 2014.07.023
33. Sperling, D., Gordon, D.: Two Billion Cars: Driving Toward Sustainability. Oxford University Press, Oxford, UK/New York (2009)
34. Andress, D., Nguyen, T.D., Das, S.: Reducing GHG emissions in the United States' transportation sector. Energy Sustain. Dev. **15**(2), 117–136 (2011)
35. Heywood, J.B.: Internal combustion engine fundamentals. In: McGraw-Hill Series in Mechanical Engineering, vol. xxix, p. 930. McGraw-Hill, New York (1988.) 2 p. of plates
36. Kim, T.Y., et al.: The effects of stratified lean combustion and exhaust gas recirculation on combustion and emission characteristics of an LPG direct injection engine. Energy. **115**, 386–396 (2016)
37. Murase, E., et al.: Initiation of combustion in lean mixtures by flame jets. Combust. Sci. Technol. **113**(1), 167–177 (2010)
38. Li, J., Yuan, L., Mongia, H.C.: Simulation of combustion characteristics in a hydrogen fuelled lean single-element direct injection combustor. Int. J. Hydrog. Energy. **42**(5), 3536–3548 (2017)
39. Rapp, V., Killingsworth, N., Therkelsen, P., Evans, R.: Chapter 4, Lean-Burn Internal Combustion Engines, pp. 111–146. Elsevier Inc (2016) https://www.sciencedirect.com/science/ book/9780128045572. ISBN: 978-0-12-804557-2
40. Zhou, F., et al.: Effects of lean combustion coupling with intake tumble on economy and emission performance of gasoline engine. Energy. **133**, 366–379 (2017)
41. Wolfhard, H.G.: The ignition of combustible mixtures by hot gases. J. Jet Propuls. **28**(12), 798–804 (1958)
42. GussakGussak, L.A.: The role of chemical activity and turbulence intensity in Prechamber-torch organization of combustion of a stationary flow of a fuel-air mixture. In: International Congress & Exposition, Detroit (1983)
43. Gussak, L., Karpov, V., Tikhonov, Y.: The Application of Lag-Process in Prechamber Engines. SAE Technical Paper 790692 (1979)
44. Gussak, L.A., M. Turkish, D. Siegla, High Chemical Activity of Incomplete Combustion Products and a Method of Prechamber Torch Ignition for Avalanche Activation of Combustion in Internal Combustion Engines. SAE Technical Paper 750890 (1975)

45. Murase, E., et al.: Initiation of combustion in lean mixtures by flame jets. Combust. Sci. Technol. **113**(1), 167–177 (1996)
46. Oppenheim, A.K.: Quest for controlled combustion engines. In: International Congress and Exposition, Detroit (1988)
47. Oppenheim, A., et al.: Jet Ignition of an Ultra-Lean Mixture. SAE Technical Paper 780637 (1978)
48. Ghoniem, A.F., Oppenheim, A.K., Chen, D.Y.: Experimental and Theoretical Study of Combustion Jet Ignition. California University, Berkeley (1983)
49. Pitt, P.L., Ridley, J.D., Clemilnts, R.M.: An ignition system for ultra lean mixtures. Combust. Sci. Technol. **35**(5–6), 277–285 (2007)
50. Yamaguchi, S., Ohiwa, N., Hasegawa, T.: Ignition and burning process in a divided chamber bomb. Combust. Flame. **59**(2), 177–187 (1985)
51. Wallesten, J., Chomiak, J.: Investigation of spark position effects in a small pre-chamber on ignition and early flame propagation. In: International Fall Fuels and Lubricants Meeting and Exposition, Baltimore (2000)
52. Elhsnawi, M., Teodorczyk, A.: Studies of mixing and ignition in hydrogen-oxygen mixture with hot inert gas injection. In: Proceedings of the European Combustion Meeting. Warsaw University of Technology ITC, Nowowiejska, Warszawa (2005)
53. Sadanandan, R., et al.: Detailed investigation of ignition by hot gas jets. Proc. Combust. Inst. **31**(1), 719–726 (2007)
54. Sadanandan, R., et al.: 2D mixture fraction studies in a hot-jet ignition configuration using NO-LIF and correlation analysis. Flow Turb. Combust. **86**(1), 45–62 (2010)
55. Gholamisheeri, M., et al.: Rapid compression machine study of a premixed, variable inlet density and flow rate, confined turbulent jet. Combust. Flame. **169**, 321–332 (2016)
56. Toulson, E., et al.: Visualization of propane and natural gas spark ignition and turbulent jet ignition combustion. SAE Int. J. Engines. **5**(4), 1821–1835 (2012)
57. Toulson, E., Watson, H., Attard, W.: Gas Assisted Jet Ignition of Ultra-Lean LPG in a Spark Ignition Engine. SAE Technical Paper 2009-01-0506 (2009)
58. Toulson, E., Watson, H., Attardm, W.: Modeling Alternative Pre-chamber Fuels in Jet Assisted Ignition of Gasoline and LPG. SAE Technical Paper 2009-01-0721 (2009)
59. Attard, W.: Turbulent jet ignition pre-chamber combustion system for spark ignition engines. MAHLE Powertrain LLC: US 20120103302 A1 (2012)
60. Attard, W.P., et al.: A New Combustion System Achieving High Drive Cycle Fuel Economy Improvements in a Modern Vehicle Powertrain. SAE Technical Paper 2011-01-0664 (2011)
61. Attard, W.P., Parsons, P.: Flame kernel development for a spark initiated pre-chamber combustion system capable of high load, high efficiency and near zero NOx emissions. SAE Int. J. Engines. **3**(2), 408–427 (2010)
62. Perera, I., Wijeyakulasuriya, S., Nalim, R.: Hot Combustion Torch Jet Ignition Delay Time for Ethylene-Air Mixtures. In: 49th AIAA Aerospace Sciences Meeting including the New Horizons Forum and Aerospace Exposition, Aerospace Sciences Meetings (2011) https://doi.org/10.2514/6.2011-95; https://arc.aiaa.org/doi/abs/10.2514/6.2011-95
63. Carpio, J., et al.: Critical radius for hot-jet ignition of hydrogen–air mixtures. Int. J. Hydrog. Energy. **38**(7), 3105–3109 (2013)
64. Karimi, A., Rajagopal, M., Nalim, R.: Traversing hot-jet ignition in a constant-volume combustor. J. Eng. Gas Turbines Power. **136**(4), 041506 (2013)
65. Shah, A., Tunestal, P., Johansson, B.: Effect of Pre-Chamber Volume and Nozzle Diameter on Pre-Chamber Ignition in Heavy Duty Natural Gas Engines, SAE Technical Paper 2015-01-0867, vol. 1, (2015) https://doi.org/10.4271/2015-01-0867. https://www.sae.org/publications/technical-papers/content/2015-01-0867/
66. Shah, A., Tunestål, P., Johansson, B.: CFD Simulations of Pre-Chamber Jets Mixing Characteristics in a Heavy Duty Natural Gas Engine. SAE Technical Paper 2015-01-1890 (2015)

Chapter 2
Ignition Mechanisms

Contents

2.1   Introduction

As described in the literature review section in the previous chapter, there exists a knowledge gap how ignition initiates by a hot turbulent jet. What are the ignition mechanisms from a fundamental point of view? What are the nondimensional parameters governing the ignition mechanism? To explore the fundamental ignition mechanisms by a hot turbulent jet, an experimental setup was built that uses a dual-chamber design (a small pre-chamber resided within the big main chamber). Two fuels, methane and hydrogen, were studied. Simultaneous high-speed schlieren and OH* chemiluminescence imaging were applied to visualize the jet penetration and ignition processes. It was found there exist two ignition mechanisms – flame ignition and jet ignition. A parametric study was conducted to understand the effects of several parameters on the ignition mechanism and probability, including orifice diameter, initial temperature and pressure, fuel/air equivalence ratios in both chambers, and pre-chamber spark position. The mean and fluctuation velocities of the transient hot jet were calculated according to the measured pressure histories in the

© Springer International Publishing AG, part of Springer Nature 2018
S. Biswas, *Physics of Turbulent Jet Ignition*, Springer Theses,
https://doi.org/10.1007/978-3-319-76243-2_2

two chambers. A limiting global Damköhler number was found for each fuel, under which the ignition probability is nearly zero. Lastly, the ignition outcome of all tests (no ignition, flame ignition, and jet ignition) was marked on the classical turbulent combustion regime diagram. These results provide important guidelines for design and optimization of efficient and reliable pre-chambers for natural gas engines.

2.2   Experimental Methods

2.2.1   Apparatus

The schematic of the experimental setup is shown in Fig. 2.1a–c. A small volume, 100 cc cylindrical stainless steel (SS316) pre-chamber was attached to the rectangular (17-inch × 6-inch × 6-inch) carbon steel (C-1144) main chamber. The main chamber to pre-chamber volume ratio was kept at 100. A stainless steel orifice plate, with a diameter ranging from 1.5 to 4.5 mm ($d = 1.5, 2.5, 3, 4.5$ mm) with a fixed orifice length to diameter ratio at $L/d = 5$, separated both chambers.

A thin, 25-micron-thick aluminum diaphragm isolated both chambers with dissimilar equivalence ratios from mixing. The fuel/air mixture in both chambers was heated up to 600 K using built-in heating cartridges (Thermal Devices, FR-E4A30TD) inserted into the main chamber walls. The mixture in the pre-chamber was ignited by an electric spark created by a 0–20 kV capacitor discharge ignition (CDI) system. An industrial grade Bosch double iridium spark plug was attached at the top of the pre-chamber. Spark location was varied using spark plugs with longer electrodes from Auburn Igniters (models I-3, I-31, I-31-2, I-32, OJ-21-5). The transient pressure histories of both chambers were recorded using high-frequency Kulite (XTEL-190) pressure transducers combined with a National Instruments' compact data acquisition chassis (C-DAQ-9178) with NI-9237 signal conditioning and pressure acquisition module via LabVIEW software. Two K-type thermocouples were positioned at the top and bottom of the main chamber to ensure uniform temperature lengthwise, thus minimizing natural convection or buoyancy effect. A 25-mm-thick polymer insulation jacket was wrapped around the pre-chamber and the main chamber to minimize heat loss. Fuel (industrial grade CH_4 and H_2) and air were introduced separately into the main chamber using the partial pressure method. Unlike the main chamber where fuel and air are mixed in the chamber, fuel/air for pre-chamber was premixed in a small stainless steel mixing chamber (2.54 cm diameter, 10 cm long) prior going into the pre-chamber.

2.2.2   Diaphragm Rupture Assessment

While the fuel/air mixture in the pre-chamber was always kept stoichiometric, the equivalence ratio in the main chamber varied from $\phi = 0.45 - 1.0$. To separate two

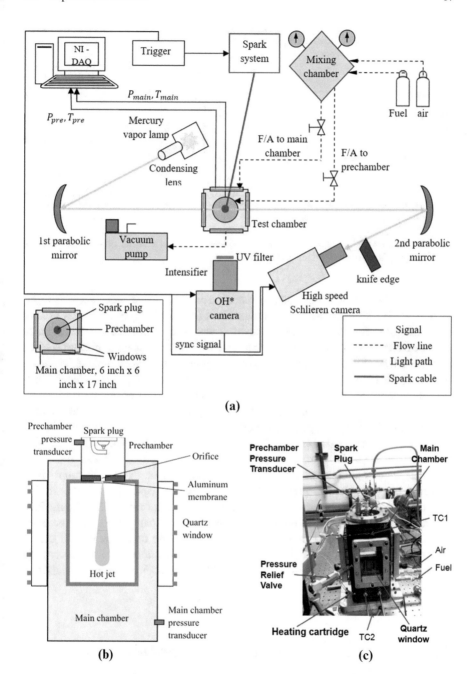

Fig. 2.1 Schematic of (**a**) the experimental setup for ignition of premixed CH_4/air and H_2/air mixtures using a hot turbulent jet generated by pre-chamber combustion, (**b**) schematic of pre-chamber and main chamber assembly, (**c**) experimental setup

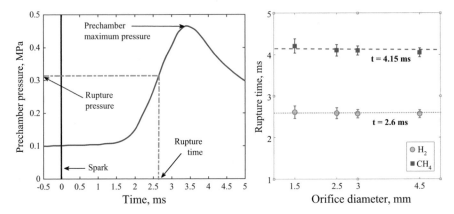

Fig. 2.2 (**a**) Rupture time and rupture pressure shown on a typical H$_2$/air pre-chamber pressure profile, (**b**) rupture time for various orifice sizes

chambers with different equivalence ratios, a diaphragm was necessary. A lightweight, 25 ± 1.25-micron-thick aluminum sheet (aluminum alloy 1100) was used as a diaphragm material. A "+"-shaped scoring was made at the diaphragm center to provide stress concentration and easy rupture. The ruptured diaphragm was replaced after each test. When the pressure difference between the two chambers reached a threshold, the diaphragm ruptured, resulting in a transient hot jet. The characterization of the threshold pressure and diaphragm rupture time was essential to accurately determine ignition delay in the main chamber. Ignition delay is defined as the time between the diaphragm rupture and the onset of ignition in the main chamber. A series of tests were conducted, and the pressure threshold was found to be 0.21 MPa which depends on the diaphragm thickness and material only. The rupture time, which is largely influenced by how fast the pressure of the pre-chamber rises, depends on the type of fuel used and the fuel/air equivalence ratio in the pre-chamber (which was fixed) only, regardless of the orifice size. Figure 2.2a shows the rupture time and rupture pressure for a typical pre-chamber pressure profile for the H$_2$/air mixture. Figure 2.2b shows rupture times for various orifice sizes for both CH$_4$/air and H$_2$/air mixtures. The CH$_4$/air mixture was observed to have a longer rupture time, 4.15 ± 0.2 milliseconds, than the H$_2$/air mixture, 2.6 ± 0.1 milliseconds.

2.2.3 High-Speed Schlieren and OH* Chemiluminescence Imaging

A customized trigger box synchronized with the CDI spark ignition system sent a master trigger to two high-speed cameras for simultaneous schlieren and OH* chemiluminescence imaging. The main chamber was installed with four rectangular (14 cm × 8.9 cm × 1.9 cm) quartz windows (type GE124) on its sides for optical access. One pair of the windows was used for the z-type schlieren system. Another pair was selected for simultaneous OH* chemiluminescence measurements. The high-speed schlieren technique was utilized to visualize the evolution of the hot jet as well as the ignition process in the main chamber. The system consisted of a 100 W (ARC HAS-150 HP) mercury lamp light source with a condensing lens, two concave parabolic mirrors (15.24 cm diameter, focal length 1.2 m), and a high-speed digital camera (Vision Research Phantom v7). Schlieren images were captured with a resolution of 800 × 720 pixels with a frame rate up to 12,000 fps.

The high-speed OH* chemiluminescence [1–3] measurement provided a better view of the ignition and flame propagation processes. A high-speed camera (Vision Research Phantom v640), along with video-scope gated image intensifier (VS4-1845HS) with 105 mm UV lens, was utilized to detect OH* signals at a very narrow band 386 ± 10 nm detection limit. The intensifier was externally synced with the camera via a high-speed relay and acquired images at the same frame rate (up to 12,000 fps) with the Phantom camera. A fixed intensifier setting (gain 65,000 and gate width 20 microseconds, aperture f8) was used all through.

2.3   Results and Discussions

As described earlier, the high-speed schlieren technique enabled visualization of the jet penetration, ignition, and subsequent turbulent flame propagation processes in the main chamber. High-speed OH* chemiluminescence was used to identify the presence of OH* radicals. It also facilitated in determining whether the hot jet coming out from the pre-chamber contains hot combustion products only or also contains active radicals such as OH. Several experiments were carried out for various initial pressures, orifice diameters, and spark locations at varying equivalence ratio, ϕ, in the main chamber for CH_4/air and H_2/air while keeping pre-chamber equivalence ratio stoichiometric. Tables 2.1 and 2.2 summarize the experimental conditions along with ignition mechanism outcomes and the ignition delay times for CH_4/air and H_2/air mixtures, respectively. Many test conditions did not result in main chamber ignition. Only the test conditions that resulted in successful main chamber ignition were included in the Tables 2.1 and 2.2. For successful ignition, two distinct mechanisms were observed: (a) flame ignition (ignition by a reacting jet, when the pre-chamber flame survived the high stretch rate and heat loss through the orifice and resulted in a jet containing many flame kernels which ignite the main chamber mixture) and (b) jet ignition (ignition by a reacted jet, when the pre-chamber flame quenched while passing through the orifice and resulted in a jet containing hot combustion products only which then ignited the main chamber mixture). These are discussed in detail in subsequent sections.

Table 2.1 Test conditions for CH₄/air ignition

Test no.	Orifice diameter (mm)	T (K)	P (MPa)	Spark location	ϕ_{pre}	ϕ_{main}	Ignition mechanism	Ignition delay, milliseconds
Effect of spark location								
1	4.5	500	0.1	Top	1.0	0.8	Jet ignition	7.82
2	4.5	500	0.1	1/3 from top	1.0	0.8	Jet ignition	7.64
3	4.5	500	0.1	Middle	1.0	0.8	Jet ignition	7.58
4	4.5	500	0.1	Bottom	1.0	0.8	Jet ignition	7.10
Effect of orifice diameter								
5	2.5	500	0.1	Top	1.0	1.0	Jet ignition	18.32
6	2.5	500	0.4	Top	1.0	1.0	Jet ignition	16.21
7	3	500	0.1	Top	1.0	1.0	Jet ignition	15.43
8	3	500	0.4	Top	1.0	1.0	Jet ignition	12.21
9	4.5	500	0.1	Top	1.0	1.0	Jet ignition	6.72
10	4.5	500	0.4	Top	1.0	1.0	Flame ignition	2.27
Effect of initial pressure								
11	4.5	500	0.1	Top	1.0	1.0	Jet ignition	6.72
12	4.5	500	0.3	Top	1.0	1.0	Flame ignition	3.08
13	4.5	500	0.4	Top	1.0	1.0	Flame ignition	2.27
Effect of main chamber equivalence ratio								
14	3	500	0.1	Top	1.0	1.0	Jet ignition	15.43
15	3	500	0.4	Top	1.0	0.9	Jet ignition	14.48
16	3	500	0.4	Top	1.0	0.8	Jet ignition	16.22
17	3	500	0.4	Top	1.0	0.7	Jet ignition	17.86
18	3	500	0.4	Top	1.0	0.5	Jet ignition	19.15
19	4.5	500	0.4	Top	1.0	1.0	Flame ignition	2.27
20	4.5	500	0.4	Top	1.0	0.9	Flame ignition	2.80
21	4.5	500	0.4	Top	1.0	0.7	Flame ignition	4.12
22	4.5	500	0.4	Top	1.0	0.6	Flame ignition	5.36
23	4.5	500	0.4	Top	1.0	0.5	Flame ignition	7.76

2.3.1   Flame Ignition (Ignition by a Reacting Jet) Mechanism

In flame ignition, the hot jet contains remnants of the pre-chamber flame. This occurs if the pre-chamber flame is not quenched by wall heat loss and high stretch rate through the orifice. Depending on the pre-chamber pressure, temperature,

Table 2.2 Test conditions for H_2/air ignition

Test no.	Orifice diameter (mm)	T (K)	P (MPa)	Spark location	ϕ_{pre}	ϕ_{main}	Ignition mechanism	Ignition delay, milliseconds
Effect of spark location								
1	4.5	300	0.4	Top	1.0	1.0	Flame ignition	1.28
2	4.5	300	0.4	1/3 from top	1.0	1.0	Flame ignition	1.22
3	4.5	300	0.4	Middle	1.0	1.0	Flame ignition	1.16
4	4.5	300	0.4	Bottom	1.0	1.0	Flame ignition	1.06
Effect of orifice diameter								
5	2.5	300	0.1	Top	1.0	1.0	Jet ignition	9.58
6	2.5	300	0.5	Top	1.0	1.0	Jet ignition	4.32
7	3	300	0.1	Top	1.0	1.0	Jet ignition	8.78
8	3	300	0.5	Top	1.0	1.0	Flame ignition	2.15
9	4.5	300	0.1	Top	1.0	1.0	Flame ignition	2.77
10	4.5	300	0.5	Top	1.0	1.0	Flame ignition	1.13
Effect of initial pressure								
11	3	300	0.1	Top	1.0	0.9	Jet ignition	9.06
12	3	300	0.3	Top	1.0	0.9	Flame ignition	3.60
13	3	300	0.5	Top	1.0	0.9	Flame ignition	2.42
Effect of main chamber equivalence ratio								
14	2.5	300	0.1	Top	1.0	1.0	Jet ignition	9.58
15	2.5	300	0.1	Top	1.0	0.9	Jet ignition	9.01
16	2.5	300	0.1	Top	1.0	0.7	Jet ignition	8.31
17	2.5	300	0.1	Top	1.0	0.6	Jet ignition	10.51
18	2.5	300	0.1	Top	1.0	0.5	Jet ignition	12.85
19	4.5	300	0.1	Top	1.0	1.0	Flame ignition	2.77
20	4.5	300	0.1	Top	1.0	0.9	Flame ignition	3.24
21	4.5	300	0.1	Top	1.0	0.7	Flame ignition	4.01
22	4.5	300	0.1	Top	1.0	0.5	Flame ignition	5.22
23	4.5	300	0.1	Top	1.0	0.45	Flame ignition	5.78

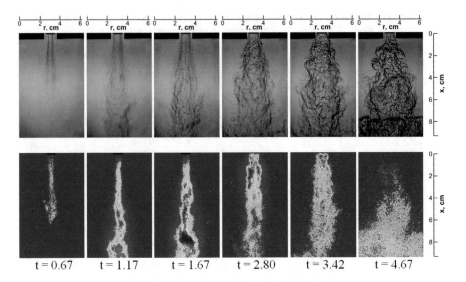

Fig. 2.3 Time sequence of simultaneous schlieren (top) and OH* chemiluminescence (bottom) images showing flame ignition process for CH$_4$/air test condition 20 in Table 2.1. $V_{pre-chamber}$ = 100 cc, $d_{orifice}$ = 4.5 mm, $P_{initial}$ = 0.4 MPa, $T_{initial}$ = 500 K, $\phi_{pre-chamber}$ = 1.0, $\phi_{main chamber}$ = 0.9, the ignition delay, $\tau_{ignition}$ is 2.80 ms

equivalence ratio, and orifice diameter, the flame that passes through the orifice can be either laminar or turbulent. For all our test conditions, it was turbulent. The hot jet contains many small turbulent flames penetrating the main chamber causing almost instantaneous ignition of the main chamber mixture.

Figures 2.3 and 2.4 show a time sequence of the simultaneous schlieren and OH* chemiluminescence images of the flame ignition processes for CH$_4$/air (test condition 20 in Table 2.1) and H$_2$/air (test condition 1 in Table 2.2), respectively. As shown in Fig. 2.3, shortly after (about 0.67 milliseconds) the diaphragm had ruptured, noticeable OH* signal was detected in the jet just coming out of the orifice from the pre-chamber. This indicates the flame in the pre-chamber did not extinguish after passing through the orifice. Rather, it developed into a turbulent flame jet. The OH* signal grew in time; a jet containing ample OH* radicals appeared at 1.67 milliseconds. The onset of the main chamber ignition occurred at 2.8 milliseconds. Ignition of the main chamber mixture was initiated from the entire surface of the jet and then propagated outwardly.

Similar behavior was observed for H$_2$/air as shown in Fig. 2.4. A noticeable amount of OH* signal was detected in the jet just coming out of the orifice, which grew over time and eventually caused ignition of the main chamber mixture at 1.28 milliseconds. For the same initial conditions and geometric configurations, H$_2$/air mixtures exhibit approximately half the ignition delay time compared to CH$_4$/air mixtures.

The pressure profiles in the pre-chamber and main chamber are shown in Fig. 2.5a and b for the CH$_4$/air and H$_2$/air cases discussed above. The trends are similar for

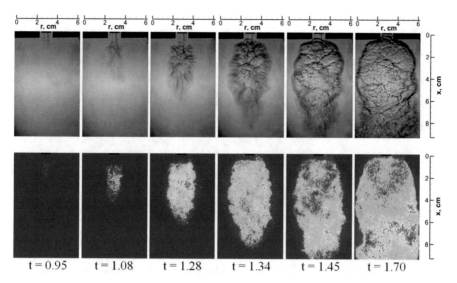

Fig. 2.4 Time sequence of simultaneous schlieren (top) and OH* chemiluminescence (bottom) images showing flame ignition process for H_2/air, test condition 1 in Table 2.2. $V_{pre-chamber} = 100$ cc, $d_{orifice} = 4.5$ mm, $P_{initial} = 0.4$ MPa, $T_{initial} = 300$ K, $\phi_{pre-chamber} = 1.0$, $\phi_{main\ chamber} = 1.0$, the ignition delay, $\tau_{ignition}$ is 1.28 ms

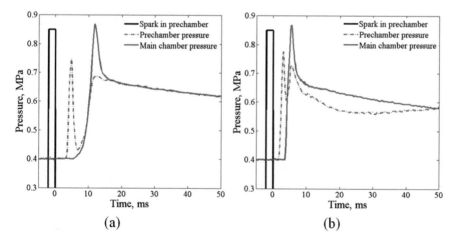

Fig. 2.5 Typical pressure profiles for flame ignition. Pressure profiles in the pre-chamber and main chamber for (**a**) CH_4/air flame ignition, test condition 20 in Table 2.1, (**b**) H_2/air flame ignition, test condition 1 in Table 2.2

both fuels, although H_2/air has shorter ignition delay. The pre-chamber pressure first raised and then reached a maximum indicating combustion in the pre-chamber was completed. Shortly after that, the pressure in the main chamber started to rise as a result of ignition in the main chamber. The main chamber pressure reached a

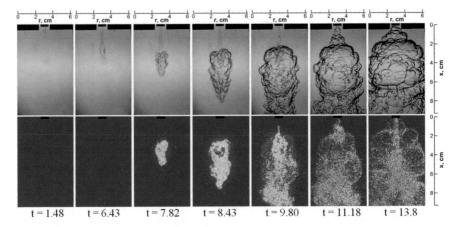

Fig. 2.6 Time sequence of simultaneous schlieren (top) and OH* chemiluminescence (bottom) images showing jet ignition process for CH_4/air, test condition 1 in Table 2.1. $V_{pre-chamber} = 100$ cc, $d_{orifice} = 4.5$ mm, $P_{initial} = 0.1$ MPa, $T_{initial} = 500K$, $\phi_{pre-chamber} = 1.0$, $\phi_{main\ chamber} = 0.8$, the ignition delay, $\tau_{ignition}$ is 7.82 ms

maximum at 12.6 milliseconds for CH_4/air and 8.9 milliseconds for H_2/air, indicating that the reactants in the main chamber were completely consumed by these times. Shortly after, the main chamber pressure started to decrease because of cooling through the chamber walls. During the combustion process of the main chamber mixture, some burned gases were pushed into the pre-chamber as the pressure in the main chamber had become higher than that of the pre-chamber. As a result, the pre-chamber pressure profile showed a second peak.

2.3.2   Jet Ignition (Ignition by a Reacted Jet) Mechanism

For jet ignition mechanism, the hot jet coming from the pre-chamber contained hot combustion products only. This means the pre-chamber flame had extinguished while passing through the orifice due to heat loss and/or high stretch rate. Because the jet contained very little or no radicals, OH* signal could not be detected at the orifice exit. Ignition by a jet of combustion products had several definite characteristics in comparison to the flame ignition mechanism and will be discussed in the following section.

Figures 2.6 and 2.7 show time sequence of simultaneous schlieren and OH* chemiluminescence images of the jet penetration and ignition processes for CH_4/air (test condition 1 in Table 2.1) and H_2/air (test condition 16 in Table 2.2), respectively. As soon as the aluminum diaphragm ruptured, the jet started flowing into the main chamber. However, unlike flame ignition mechanism, no appreciable OH* chemiluminescence signal was detected in the jet coming out of the orifice during this period. This indicates the pre-chamber flame had been quenched when passing

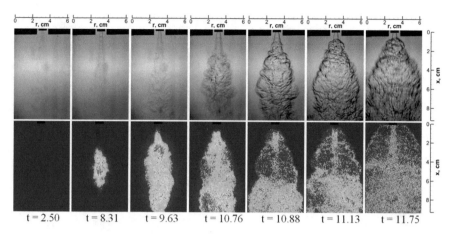

Fig. 2.7 Time sequence of simultaneous schlieren (top) and OH* chemiluminescence (bottom) images showing jet ignition process for H_2/air, test condition 16 in Table 2.2. $V_{pre-chamber} = 100$ cc, $d_{orifice} = 2.5$ mm, $P_{initial} = 0.1$ MPa, $T_{initial} = 300$ K, $\phi_{pre-chamber} = 1.0$, $\phi_{main\ chamber} = 0.7$, the ignition delay, $\tau_{ignition}$ is 8.31 ms

through the orifice. After some time, OH* signal was first detected at a location a few centimeters downstream of the orifice, and the main chamber pressure started to rise. The jet lasted about 7.82 milliseconds for CH_4/air and 8.31 milliseconds for H_2/air before it ignited the main chamber mixture. These ignition delay times are longer in comparison to the typical ignition delays observed in the flame ignition (H_2/air $\sim$ 1–6 milliseconds, CH_4/air $\sim$ 1–8 milliseconds).

Additionally, we found that ignition started from the side surface of the jet. Ignition at the jet tip was not observed for any of our test conditions where the jet ignition mechanism holds. This is consistent with the observations of Elhsnawi [4] but conflicts with Sadanandan [5, 6] who observed ignition starting at the jet tip. Furthermore, both the schlieren and OH* chemiluminescence images showed that ignition started in a region that was 2–3 cm downstream of the orifice exit for CH_4/air and 4–5 cm downstream of orifice exit for H_2/air. The reason why ignition took place in these regions/locations is explained by examining the jet velocities and the Damköhler numbers in the following sections.

Figure 2.8a and b show the typical pressure histories in the pre-chamber and the main chamber for the CH_4/air and H_2/air cases discussed above. The ignition delay for CH_4/air was similar to that of H_2/air. After the onset of spark ignition in the pre-chamber, its pressure started to rise. This increased pressure in the pre-chamber and created a transient pressure difference responsible for driving the hot combustion products through the orifice in the form of a turbulent hot jet. At the end of main chamber combustion, which is marked by the peak of the main chamber pressure profile, part of the main chamber combustion products entered into the pre-chamber. Therefore, the pre-chamber pressure increased again following the main chamber pressure profile.

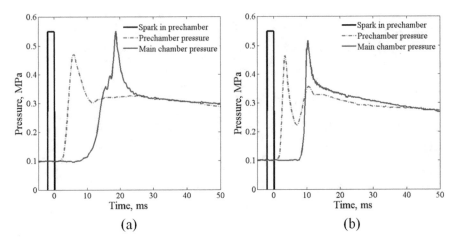

Fig. 2.8 Typical pressure profiles for jet ignition. Pressure profiles in the pre-chamber and main chamber for (**a**) CH_4 jet ignition, test condition 1 in Table 2.1, (**b**) H_2 jet ignition, test condition 16 in Table 2.2

2.3.3   Parametric Effects on Ignition Mechanisms and Ignition Probability

The two ignition mechanisms discussed in preceding sections have certain identifiable characteristics. The ignition delay time for flame ignition is rather short as compared to jet ignition. This is because the flame jet contains key radicals and intermediate species that promote chain-branching reactions. On the contrary, jet ignition takes longer time because the quenched flames need time to mix with the cold unburned ambient gases in the main chamber. Furthermore, flame ignition was normally initiated from the entire surface of the jet, whereas jet ignition was always observed starting from the lateral sides of the jet at a location a few centimeters downstream of the orifice. Lastly, for jet ignition, OH* signal was absented in the jet coming out of the pre-chamber through the orifice, while it can be detected in the jet for flame ignition. For instance, both Figs. 2.6 and 2.7 suggest that the pre-chamber flame had quenched in the orifice and the hot jet consisted of combustion products only. The occurrence of flame or jet ignition mechanism depends on a number of parameters such as orifice diameter, initial pressure, fuel type, and equivalence ratios. In the following sections, we will discuss the effects of these parameters on ignition mechanisms and ignition probability.

The results presented in Tables 2.1 and 2.2 show that keeping all other variables unchanged, a decreasing orifice diameter switches the ignition mechanism from flame ignition to jet ignition. A smaller orifice imposes a higher stretch rate to the pre-chamber flame. Hence flame extinction becomes more likely in a smaller orifice leading to jet ignition. This observation is consistent with the results of Iida [7] who studied the transient behavior of CH_4/air flame flowing into a narrow channel at atmospheric pressure. They found that the flame can either pass, standstill, or extinguish depending on the mixture equivalence ratio, the channel width, and the flame inflow velocity. They also found a critical channel width of 2–3.5 mm, below which the flame could not survive and would extinguish due to high stretch rate which consequently resulted in high mixing rate between the flame front and the cold unburned ambient gases.

The results in the Tables 2.1 and 2.2 also suggest that as pressure increases flame ignition mechanism becomes more prevalent. This attribute can well be explained by flame quenching behavior as a function of pressure. As defined in [8], the quenching distance, d_q, is the minimum separation distance between two cold, flat plates, beyond which a flame cannot pass. Quenching distance is expected to be on the same order as flame thickness. This is because the quenching distance is the characteristic length through which heat is conducted from the hot flame to the cold wall. It is reasonable to assume that quenching distance is proportional to laminar flame thickness, $d_q \sim l_f$ [8]. With the increase in pressure, CH_4/air and H_2/air flame thicknesses decrease. As a result, a higher pressure resulted in a shorter quenching distance. This means that at high pressures, pre-chamber flame can pass through a smaller diameter orifice without quenching. Thus, ignition mechanism shifted to flame ignition at higher pressures.

For all test conditions, the pre-chamber mixture was kept at stoichiometry, $\phi = 1$, whereas the main chamber mixture was varied from stoichiometric to lean conditions. Decreasing the main chamber equivalence ratio will reduce ignition probability, although it does not affect the ignition mechanism. Lastly, to understand the effect of spark location in the pre-chamber on the ignition mechanism and probability, three locations were tested (one very close to the top of the pre-chamber, one in the middle, and the third one very close to the bottom of the pre-chamber and near the orifice). The results showed that spark location had a negligible effect on ignition mechanism and probability. However, depending on the spark plug position, the ignition delay time varied slightly. The spark plug location closest to the orifice resulted in marginally shorter (10–15% lower) ignition delay times as compared to the farthest spark location.

2.3.4 Jet Characteristics and the Global Damköhler Number

To understand the jet ignition mechanism and the complex interactions between chemistry and turbulent mixing, knowledge of the hot transient jet produced by pre-chamber combustion is required. For a given orifice diameter, d, the mean jet velocity at the orifice exit, U_0, was calculated based on the experimentally measured pressure drop between the pre-chamber and the main chamber, $\Delta P(t) = P_{pre}(t) - P_{main}(t)$, and the gas properties. Due to the low viscosity of pre-chamber combustion products (order of $\sim O[10^{-5}]$), the boundary layer thickness is negligible compared to orifice diameter, d. Thus, the velocity profile at the orifice exit can be assumed to be flat-topped (plug flow). Then the mean jet velocity of the compressible gas flow at the orifice exit, $U_0(t)$, can be expressed as [9]

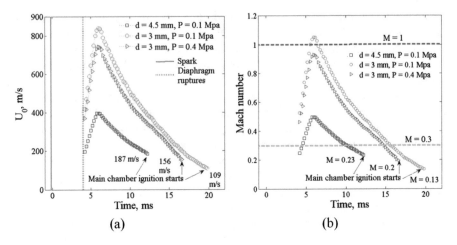

Fig. 2.9 (a) The mean centerline jet velocity at the orifice exit, U_0, as a function of time for various CH$_4$/air test conditions. (b) Mach number as a function of time for the identical test conditions as shown in (a)

$$U_0(t) = CY\sqrt{\frac{2\Delta P(t)}{\rho_{\text{mix}}(t)}} \tag{2.1}$$

where $\rho_{\text{mix}}(t) = \sum_i X_i \rho_i$ is the density of the pre-chamber gas mixture, $C = C_d/\sqrt{1-\beta^4}$, and Y is the expansion factor to account for the compressibility effect, and it is a function of γ, β and $p_{\text{pre}}/p_{\text{main}}$. Y was then calculated using the tabulated data presented in [9]. Heat loss through the orifice wall was neglected. Even though the orifice diameters were small, the transient jet lasted for 1–20 milliseconds only. This small amount of time coupled with the high velocity of the jet while passing through the orifice did not result in significant heat conduction to the orifice wall. The jet exit temperatures were measured using hot wire pyrometry technique and showed 2.5% variation of the hot jet temperature from calculated temperatures without considering heat loss. Additionally, the Kulite pressure transducer had an uncertainty $\pm 0.1\%$ of its full-scale value. This resulted in an uncertainty of 2.51% in density and 2.92% in velocity calculations due to uncertainty in pressure and temperature.

Figure 2.9a shows time evolution of the mean centerline jet velocity at the orifice exit, U_0, for three different CH$_4$/air test conditions (test numbers 1, 7, and 8, respectively). All of them involved the jet ignition mechanism. The purpose was to show the effect of orifice diameter, d, and the initial pressure, p, on the jet velocities and subsequently on the ignition mechanism. As expected, for a given pressure, a smaller orifice resulted in higher jet velocities than a larger orifice. For a fixed orifice size, as pressure goes up, the density increased significantly. However, at higher pressure the pressure difference, ΔP, increased as well. Thus, for a fixed diameter orifice, the centerline mean jet velocity varies only slightly in the pressure range of

0.1–0.4 MPa, as shown in (a). For all cases, the hot jet started to flow into the main chamber after the diaphragm ruptured 4.15 milliseconds after spark ignition in the pre-chamber. As the pre-chamber pressure went up, the jet accelerated, reached a maximum, and then started decelerating. We marked the instant at which ignition of the main chamber mixture took place in Fig. 2.9a. The results show that main chamber ignition did not take place during the jet acceleration process. Rather, it always occurred when the jet was decelerating. The corresponding centerline velocity U_0 at the orifice exit for these three cases ranges from 109 m/s to 187 m/s. These velocities were less than half of their respective maximums.

To evaluate the compressibility effect, the Mach numbers of the CH_4/air jets were plotted. Figure 2.9b shows the Mach number as a function of time corresponding to the transient velocity profiles discussed in Fig. 2.9b. Only for a small period of time during its lifetime, the hot jet became choked. Although for the majority of its lifetime, it remained subsonic. We observed that the main chamber ignition started at a Mach number that was much lower than the maximum Mach number of the jet during its lifetime. Although the maximum Mach number for CH_4/air was ranging from 1.4 to 0.41, for the majority of the test conditions, the maximum Mach number remained subsonic. The ignition Mach number indicated that the jet exit Mach number, just prior to the ignition in the main chamber, remained less than 0.3 for all test conditions.

As the hot jet penetrates into the main chamber, the jet surface contains many small eddies. These eddies help to mix the hot jet with the cold, unburned fuel/air mixture in the main chamber. As turbulent eddies dissipate energy, the temperature of the hot jet drops during the penetrating process. If the jet temperature drops too rapidly, it may not be able to ignite the main chamber mixture. The competition between the turbulent mixing timescale and the chemical timescale, characterized by the Damköhler number, has a deterministic effect on the ignition outcome (successful or failed). The maximum initial pressure and temperature in the present experiments were limited to 0.5 MPa and 500 K. However, the prior ignition pressure and temperature in natural gas engines can be as high as 10–25 MPa and 800–900 K. Thus, it was important to remove the parametric dependency of current data so that the knowledge developed here can be applied to engine relevant conditions. To make the results more useful, we generated a diagram to illustrate the range of the Damköhler numbers that likely results in the successful ignition (high ignition probability) based on the many tests we had performed. This diagram could be used as a tool to evaluate ignition probability at various operating and design conditions such as different pressures, temperatures, equivalence ratios, and pre-chamber designs.

The Damköhler number, Da, was defined as the ratio of the characteristic flow timescale, τ_F, to the characteristic chemical reaction timescale, τ_C:

$$Da = \frac{\tau_F}{\tau_C} \tag{2.2}$$

Table 2.3 Global Damköhler number calculation for CH_4/air at 1 atm and 300 K, orifice diameter is 2.5 mm

ϕ	s_L, cm/s	l_F, µm	u', m/s	Da
0.6	10.52	27	1.24	39
0.7	19.34	21	1.62	70
0.8	27.91	17	1.78	111
0.9	32.65	14	1.93	145
1.0	37.41	12	2.42	155

Table 2.4 Global Damköhler number calculation for H_2/air at 1 atm and 300 K, orifice diameter is 2.5 mm

ϕ	s_L, cm/s	l_F, µm	u', m/s	Da
0.5	54.45	16	5.12	43
0.6	90.32	12.5	7.63	56
0.7	121.1	10	11.21	58
0.8	148.94	8	16.25	59
0.9	175.32	6	20.18	72
1.0	201.53	4.5	22.72	93

Here the characteristic flow timescale is the turbulent mixing timescale, which largely depends on the turbulent mean and fluctuation velocities (or the turbulent Reynolds number). The characteristic chemical timescale is the ignition timescale, which mainly depends on the chemistry of the reactions and the temperature at which the reactions take place. The Damköhler number can further be written in terms of the fuel/air thermophysical properties and flow field information as [10]

$$Da = \frac{s_L l}{u' l_f} \tag{2.3}$$

s_L and l_f were calculated using the PREMIX module of ChemkinPro [11]. l_f was estimated as the full width at half maximum (FWHM) of the volumetric heat release rate from the PREMIX calculations. The local velocity fluctuations at the orifice exit were determined using I and U_0, $u' = U_0 I$. Turbulent intensity for internal flows can be estimated using an empirical correlation [12], $I = u'/U_0 = 0.16 Re_d^{-1/8}$. The integral length scale was estimated along the orifice lip line using the correlation for compressible round jets [13], $l/d = 0.052 x/d + 0.0145$, where x is the distance downstream of orifice exit. Tables 2.3 and 2.4 provide values for turbulent intensity, laminar flame speed, and flame thickness to calculate the global Damköhler number for CH_4/air and H_2/air at 1 atm and 300 K, using an orifice diameter of 2.5 mm.

Ignition is a highly localized phenomenon. Ideally, the Damköhler number near the ignition location, i.e., "local" Damköhler number, should be calculated considering "local" flow conditions. However, such information is very difficult to obtain experimentally because of the unsteady, highly transient nature of the hot jet. So, we defined a "global" Damköhler number using the orifice exit as the location and at the time just prior to the main chamber ignition. For each test condition that results in a successful ignition (either by flame ignition or by jet ignition), we calculated the global Damköhler number based on the jet mean and fluctuation velocity at the orifice exit at the time of main chamber ignition, as well as the laminar flame speed

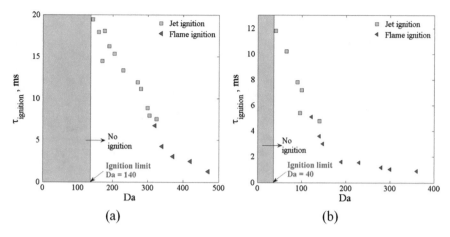

Fig. 2.10 Ignition delay, $\tau_{ignition}$, as a function of the global Damköhler number, Da, for all test conditions involving (**a**) CH$_4$/air and (**b**) H$_2$/air

and the flame thickness, both of which were obtained from PREMIX calculations. This "global" Damköhler number would give us insights on the competition between turbulent mixing and ignition chemistry as well as understanding of the effect of a number of parameters such as orifice diameter on ignition probability.

We plotted the ignition delay time as a function of the global Damköhler number for all test conditions summarized in Tables 2.1 and 2.2, as shown in Fig. 2.10a for CH$_4$/air and in Fig. 2.10b for H$_2$/air.

The range of Damköhler numbers under which successful ignition takes place is Da = 140 to 500 for CH$_4$/air and Da = 40 to 380 for H$_2$/air. The lower limit (Da$_{crit}$ = 140 for CH$_4$/air and Da$_{crit}$ = 40 for H$_2$/air) separating no-ignition and ignition regimes was referred to as the critical Damköhler number, Da$_{crit}$. Below Da$_{crit}$, turbulent mixing timescale was too small as compared to the chemical timescale, indicating rapid mixing and heat loss and thus extinction of the ignition kernels. Moreover, the flame ignition cases had lower ignition delay time and higher Damköhler numbers than jet ignition cases. Flame and jet ignition mechanisms were divided by a Damköhler number of 300~350 for CH$_4$/air and 120~130 for H$_2$/air. These are referred to as transition Damköhler numbers, Da$_{tran}$.

Critical conditions necessary for ignition were considerably influenced by the Lewis number of the reactant mixture. Lewis number, Le, is defined as thermal diffusivity of the mixture to the mass diffusivity of the deficient reactant. Le was smaller than unity (0.2 < Le < 0.6) for fuel-lean H$_2$/air mixtures and near unity for CH$_4$/air mixtures. Our results showed that the effect of higher reactant diffusivity of H$_2$ lowered the critical Damköhler number ($D_{critical}$ = 40 for H$_2$ as compared to $D_{critical}$ = 140 for CH$_4$). This observation was consistent with the Iglesias et al.'s numerical study on the effect of Lewis number on initiation of deflagration by a hot jet [14]. As reflected by smaller values of the fuel Lewis number for H$_2$, the cold unburned H$_2$/air rapidly diffused into the hot jet before the jet loses much of its heat

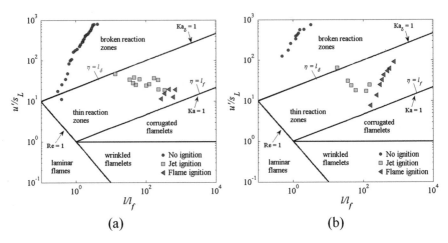

Fig. 2.11 The ignition outcome of all test conditions involving (**a**) CH$_4$/air and (**b**) H$_2$/air are presented in the premixed turbulent combustion regime diagram. The diagram is based on Peters [10]

to the cold surroundings. This preheated H$_2$/air facilitates favorable combustion conditions, and combustion becomes possible at lower Damköhler number than for CH$_4$/air, as has been explained in [14, 15].

2.3.5   Turbulent Premixed Combustion Regimes for Hot Jet Ignition

Diagrams defining regimes of premixed turbulent combustion in terms of velocity and length scale ratios have been proposed by Borghi [16] and later modified by Peters [17], Abdel-Gayed [18], Poinsot [19], Williams [20], and others. As shown in Fig. 2.11a and b, several regimes exist including laminar flames, turbulent flamelets, thin reaction zones, and broken reaction zones based on the logarithm of velocity ratio, u'/s_L, and length scale ratio, l/l_f. They indicate different interactions between turbulence and chemistry. It would be interesting to see how the no-ignition, jet ignition, and flame ignition cases fitted in the premixed turbulent combustion regime diagram. Motivated by this, the ignition outcome for each test condition in Tables 2.1 and 2.2 along with no-ignition cases was plotted in the turbulent premixed combustion diagram, as shown in Fig. 2.11a for CH$_4$/air and in Fig. 2.11b for H$_2$/air.

Most no-ignition cases fall into the broken reaction zones for both CH$_4$/air and H$_2$/air. Mansour and Chen thoroughly investigated the stretched premixed flames on a Bunsen burner that were surrounded by a large pilot flame in a series of studies [21, 22]. Their results show that in the broken reaction zone regime, a premixed flame is unable to survive. The present experimental observations are consistent with this conclusion, as local extinction events would appear more frequently when flames enter into the broken reaction zones regime. The no-ignition cases corresponded to pre-chambers with smaller orifice diameters, at lower initial temperatures, higher initial pressures, and fuel-lean conditions. The fundamental reason for no ignition is that the turbulent jet is very strong with relatively high mean velocity and fluctuations. This results in rapid mixing between the hot jet and the cold unburned main chamber mixture. The initial temperature of the unburned mixture, however, was limited to a maximum of 500 K in the present experiments. These result in relatively large chemical timescales and thicker preheat zones and reaction zones in the flame. As a result, the smallest turbulent eddies are smaller than the inner reaction zone thickness. Turbulent eddies can penetrate into the inner reaction zone perturbing it and causing local chemistry to break down because of increased heat loss to the preheat zone [10].

Most ignition cases fall within the thin reaction zone regime for both CH_4/air and H_2/air, with a few at the boundary between the thin reaction zones regimes and the broken reaction zones regime. None of them fall within the corrugated flamelet regime. In general, flame ignition cases correspond to a larger l/l_f ratio. As we increase orifice diameter and pressure, the ignition mechanism tends to switch from jet ignition to flame ignition. The flame jet coming out the pre-chamber orifice contains key radicals such as H, O, and OH, which initiate chain-branching reactions and enhance ignition probability. Compared to CH_4/air, H_2/air flames have higher burning velocities and thinner preheat and reaction zones. In other words, the chemical timescale for H_2/air flames was shorter than CH_4/air flames. As a result, under the same level of turbulence, H_2/air has a better chance for ignition than CH_4/air, even at lower initial temperatures.

2.4 Conclusions

We investigated the ignition characteristics of CH_4/air and H_2/air mixtures using a turbulent hot jet generated by pre-chamber combustion using simultaneous high-speed schlieren and OH* chemiluminescence. The vital finding was the existence of two different ignition mechanisms, namely, jet ignition and flame ignition. The former produced a jet of hot combustion products (which means the pre-chamber flame is quenched when passing through the orifice); the latter produced a jet of wrinkled turbulent flames (the composition of the jet is incomplete combustion products containing flames). As the orifice diameter increased, the ignition mechanism tends to switch to flame ignition, from jet ignition. With the increase in pressure, flame ignition became more prevalent. The ignition took place on the side surface of the hot jet during the jet deceleration process for both mixtures. A critical global Damköhler number, Da_{crit}, defined as the limiting parameter that separated ignition from no ignition, was found to be 140 for CH_4/air and 40 for H_2/air. All possible ignition outcomes were plotted on the turbulent combustion regime diagram. Nearly all no-ignition cases fell into the broken reaction zone, and jet and flame ignition cases mostly fell within the thin reaction zones.

References

1. Smith, G.P., et al.: Low pressure flame determinations of rate constants for OH(A) and CH (A) chemiluminescence. Combust. Flame. **131**(1), 59–69 (2002)
2. Luque, J., et al.: CH(A-X) and OH(A-X) optical emission in an axisymmetric laminar diffusion flame. Combust. Flame. **122**(1), 172–175 (2000)
3. Orain, M., Hardalupas, Y.: Measurements of local mixture fraction of reacting mixture in swirl-stabilised natural gas-fuelled burners. Appl. Phys. B. **105**(2), 435–449 (2011)
4. Elhsnawi, M., Teodorczyk, A.: Studies of mixing and ignition in hydrogen-oxygen mixture with hot inert gas injection. In: Proceedings of the European Combustion Meeting. Warsaw University of Technology ITC, Nowowiejska, Warszawa (2005)
5. Sadanandan, R., et al.: 2D mixture fraction studies in a hot-jet ignition configuration using NO-LIF and correlation analysis. Flow Turb. Combust. **86**(1), 45–62 (2010)
6. Sadanandan, R., et al.: Detailed investigation of ignition by hot gas jets. Proc. Combust. Inst. **31** (1), 719–726 (2007)
7. Iida, N., Kawaguchi, O., Sato, G.T.: Premixed flame propagating into a narrow channel at a high speed, part 2: transient behavior of the properties of the flowing gas inside the channel. Combust. Flame. **60**(3), 257–267 (1985)
8. Law, C.K.: Combustion Physics, vol. xviii, p. 722. Cambridge University Press, Cambridge, MA/New York (2006)
9. Crane Co: Engineering Division. In: Flow of Fluids through Valves, Fittings, and Pipe. Crane Company Technical paper. Crane Co, Chicago (1957)
10. Peters, N.: Turbulent combustion. In: Cambridge Monographs on Mechanics, vol. xvi, p. 304. Cambridge University Press, Cambridge, MA/New York (2000)
11. Reaction Design: Reaction Workbench 15131 San Diego (2013) http://www.reactiondesign. com/support/help/help_usage_and_support/how-to-citeproducts/
12. Debonis, J.R., Scott, J.N.: Large-Eddy simulation of a turbulent compressible round jet. AIAA J. **40**(7), 1346–1354 (2002)
13. Uzun, A., Hussaini, M.Y.: Investigation of high frequency noise generation in the near-nozzle region of a jet using large eddy simulation. Theor. Comput. Fluid Dyn. **21**(4), 291–321 (2007)
14. Iglesias, I., et al.: Numerical analyses of deflagration initiation by a hot jet. Combust. Theory Modell. **16**(6), 994–1010 (2012)
15. Carpio, J., et al.: Critical radius for hot-jet ignition of hydrogen–air mixtures. Int. J. Hydrog. Energy. **38**(7), 3105–3109 (2013)
16. Borghi, R.P.: On the structure and morphology of turbulent premixed flames. In: Casci, C. (ed.) Recent Advances in the Aerospace Sciences, pp. 117–138. Springer, Boston (1985)
17. Peters, N.: Laminar flamelet concepts in turbulent combustion. Int. Symp. Combust. **21**(1), 1231–1250 (1988)
18. Abdel-Gayed, R.G., Bradley, D., Lung, F.K.K.: Combustion regimes and the straining of turbulent premixed flames. Combust. Flame. **76**(2), 213–218 (1989)
19. Poinsot, T., Veynante, D., Candel, S.: Diagrams of premixed turbulent combustion based on direct simulation. Int. Symp. Combust. **23**(1), 613–619 (1991)
20. Williams, F.A.: Combustion theory: the fundamental theory of chemically reacting flow systems. In: Combustion Science and Engineering Series, vol. xxiii, 2nd edn, p. 680. Benjamin/Cummings Pub. Co, Menlo Park (1985)
21. Chen, Y.C., et al.: The detailed flame structure of highly stretched turbulent premixed methane-air flames. Combust. Flame. **107**(3), 223–244 (1996)
22. Chen, Y.C., Mansour, M.S.: Measurements of the detailed flame structure in turbulent H2-Ar jet diffusion flames with line-Raman/Rayleigh/LIPF-OH technique. Int. Symp. Combust. **26**(1), 97–103 (1996)

Chapter 3
Schlieren Image Velocimetry (SIV)

Contents

3.1 Introduction

Particle image velocimetry (PIV) is a quantitative optical method used in experimental fluid dynamics that captures entire 2D/3D velocity field by measuring the displacements of numerous small particles that follow the motion of the fluid. In its simplest form, PIV acquires two consecutive images (with a very small time delay) of flow field seeded by these tracer particles, and the particle images are then cross-correlated to yield the instantaneous fluid velocity field. The nature of PIV measurement is rather indirect as it determines the particle velocity instead of the fluid velocity. It is assumed in PIV that tracer particles "faithfully" follow the flow field without changing the flow dynamics. To achieve this, the particle response time should be faster than the smallest time scale in the flow. The flow tracer fidelity in PIV is characterized using Stokes number, S_k, where a smaller Stokes number ($S_k < 0.1$) represents excellent tracking accuracy. Conversely, schlieren and shadowgraph are truly nonintrusive techniques that rely on the fact that the change in

© Springer International Publishing AG, part of Springer Nature 2018
S. Biswas, *Physics of Turbulent Jet Ignition*, Springer Theses,
https://doi.org/10.1007/978-3-319-76243-2_3

refractive index causes light to deviate due to optical inhomogeneities present in the medium. Schlieren methods can be used for a broad range of high-speed turbulent flows containing refractive index gradients in the form of identifiable and distinguishable flow structures. In schlieren image velocimetry (SIV) techniques, the eddies in a turbulent flow field serve as PIV "particles." Unlike PIV, there are no seeding particles in SIV. To avoid confusion, a quotation mark is used for "particles" when describing the SIV techniques. As the eddy length scale decreases with the increasing Reynolds number, the length scales of the turbulent eddies become exceptionally important. These self-seeded successive schlieren images with a small time delay between them can be correlated to find velocity field information. Thus, the analysis of schlieren and shadowgraph images is of great importance in the field of fluid mechanics since this system enables the visualization and flow field calculation of unseeded flow.

Papamoschou [1] showed the possibility of using schlieren technique to measure the velocity of very high-velocity flows. A schlieren system with a pulsed light source and a very short (~20 ns) camera exposure time was utilized to measure a supersonic shear layer. However, due to lack of processing power, only a global convective velocity of the turbulent shear layer was captured using pattern matching technique between two consecutive schlieren images. Fu [2, 3] was among the first to use high-speed (~100 kHz) imaging to extract velocity field from sequences of schlieren images. All previous works used application-specific image-processing algorithm, until the substantial development of digital PIV (DPIV) toward the end of the twentieth century, when the digital PIV processing algorithm was commonly and commercially available.

Raffel [4–6], Elsinga [7, 8], and Scarano [9] computed flow field information using background oriented schlieren (BOS) technique using the evaluation algorithms proposed by Rafael and Goldhahn [10, 11]. Kegerise and Settles performed image correlation velocimetry on an axisymmetric turbulent-free convection plume [12]. Garg and Settles performed turbulence measurements of the supersonic turbulent boundary layer by focusing schlieren deflectometry technique [13]. Jonassen and Settles [14] explored the possibility of using commercial PIV equipment combined with schlieren optics to measure the velocity field of an axisymmetric helium jet and a 2D supersonic turbulent boundary layer. A laminar jet structure at the exit prevented them from capturing the tip and near-field velocities of helium jet. However, Jonassen compared different schlieren and shadowgraph sensitivity settings to find an optimum sensitivity value. Later Hargather et al. [15, 16] compared three quantitative schlieren techniques, Schardin's calibrated schlieren, rainbow schlieren, and background oriented schlieren (BOS) using a 2D turbulent boundary layer. Most recently Mauger [17] performed velocity measurements in a cavitating microchannel two-phase flow configuration using shadowgraph image correlations. Zelenak [18] conducted an experiment to test the applicability of the laser shadowgraph technique combined with commercial PIV processing algorithms to visualize the pulsating water jet structure and to analyze the velocity field.

All the aforementioned schlieren or schlieren-like velocimetry techniques used either application-specific, homegrown image processing algorithms or commercial digital PIV software to obtain flow field information. Since commercial PIV software had been developed keeping digital PIV in mind, processing parameters are not optimized for SIV.

Until now all studies on schlieren velocimetry techniques were accentuated how well SIV techniques agree compared to a well-established velocity measurement technique such as hot wire, LDV, or PIV. No such detailed study was conducted on the performance of different image correlation algorithms and the effect of various preprocessing and post-processing techniques for SIV. Additionally, very few past studies used SIV to resolve high Reynolds number flows except Hargather's work [16] in which the mean velocity profile of a Mach 3 turbulent boundary layer was measured. The challenge for high-speed flows lies in the fact that SIV requires extremely expensive digital cameras – either a very high-speed camera (100 kHz or up) or a multi-exposure camera is needed to resolve a high-speed velocity field using SIV.

Motivated by this, the present study provides a comprehensive statistical assessment of three different SIV techniques, schlieren with horizontal knife-edge cutoff, schlieren with vertical knife-edge cutoff, and shadowgraph, for a high-velocity (0.3 < Mach < 0.6) helium jet using an open source robust phase correlation (RPC) [19] code. RPC shows enhanced measurement capabilities regarding better signal to noise (SNR) ratio, reduction of bias error, and peak locking. This study is one of the very first to apply SIV techniques to measuring high-speed flows. Experimental measurement of high-speed flows is extremely challenging. At high speeds, the chances that PIV tracer particles will not "faithfully" follow the flow increase. Thus, it becomes increasingly difficult to apply PIV to high-speed flows. Since SIV techniques do not require seeding particles, they provide an alternate solution. However, to date, very few studies measured velocities of high-speed flows using SIV techniques due to two reasons. The first reason is hardware limitations of the camera, and the second is due to the absence of a well-established post-processing routine. In the present study, we proposed a novel, inexpensive two-camera approach to solving this problem. Using two high-speed cameras side by side to capture schlieren images effectively forms a double-exposure high-speed camera. This two-camera approach has never been explored and can be very helpful to resolve high-speed axisymmetric or 2D flows. A highly resolved temporal and spatial velocity field was obtained using two cameras in parallel. Unlike PIV, image preprocessing, conditioning, filtering, image processing, and post-processing are all different for SIV. Optimization of SIV processing parameters and various types of image filtering, image restoration, and noise reduction techniques useful for SIV techniques were discussed in detail. Because the ratio of the time scales of the smallest and largest eddies (τ_η and τ_0) varies with Reynolds number, $\tau_\eta/\tau_0 \sim Re^{-1/2}$ [20], the size ratio of the biggest and smallest windows in SIV methods also depends on the Reynolds number. Although previous studies compared their SIV results to PIV, LDV, or hot-wire measurements, none statistically explained the performance of SIV. A comprehensive analysis of different SIV techniques was carried out using the correlation plane statistics. Correlation plane statistics for various SIV techniques depended primarily on turbulence parameters such as turbulence intensity.

Quantitative comparison of correlation planes at different spatial locations of high-speed axisymmetric helium jet was performed. Several performance metrics such as primary peak ratio (PPR), peak to correlation energy (PCE), and the probability distribution of SNR were used to compare capabilities of different SIV techniques. This study combined an open source PIV processing algorithm with a novel two-camera, easy to setup SIV technique with a detailed description of image preprocessing, flow field post-processing, and their statistical assessment, therefore presented a solution to resolve velocity fields of a wide range of turbulent flows.

3.2   Experimental Methods

The experimental setup is schematically shown in Fig. 3.1a. A steady helium jet was issued from a nozzle at velocity U_0 into a quiescent ambient. The nozzle had a diameter of 4 mm and a length to diameter ratio, L/d of 2. Schematic of the axisymmetric jet coordinates (x, r) is shown in Fig. 3.1b. The jet centerline velocity $U_0(x)$ decayed as the jet spread $\delta(x)$ in the streamwise direction. The upstream nozzle pressure was carefully controlled using Kulite (XTEL-190) pressure transducers combined with NI-9237 signal conditioning module to create two different flow conditions at the nozzle exit: $Re_d = 11,000$ and $Re_d = 22,000$, respectively. The jet Reynolds number, Re_d, is defined as $Re_d = U_0 d/\nu$, where U_0 is the jet exit velocity, d is the nozzle diameter, and ν is the kinematic viscosity at the nozzle exit. For Reynolds number 11,000 and 22,000, the jet centerline velocity at the nozzle exit were 304 m/s (Mach = 0.3) and 611 m/s (Mach = 0.6), respectively, based on the PIV measurements. The theoretical jet exit velocity calculated from the experimentally measured pressure drop across the nozzle using plug flow assumptions were 297 m/s and 601 m/s, respectively, which agreed well with the PIV data.

3.2.1   High-Speed Schlieren and Shadowgraph Imaging

A z-type Herschellian high-speed schlieren system was used to visualize the axisymmetric turbulent helium jet. The schlieren system consisted of a 100 Watt mercury arc lamp (Q series, 60,064-100MC-Q1, Newport Corporation, Model 6281) light source with a condensing lens assembly (Q Series, F/1, Fused Silica, Collimated, 200–2500 nm), two concave parabolic mirrors (6″ diameter, aperture $f/8$, effective focal length 1219.2 mm), a knife-edge, an achromatic lens ($f = 300$ mm) to collimate the light, a beam splitter (1″ cube, Thorlabs PBS251), and two identical high-speed CC-D cameras (Phantom v711, Vision Research). The inter-frame delay, Δt, between the two high-speed cameras was accurately controlled by a low jitter digital delay generator (Stanford Research, DG535). The inter-frame delay is needed to be small enough to resolve high jet velocity. The same setup

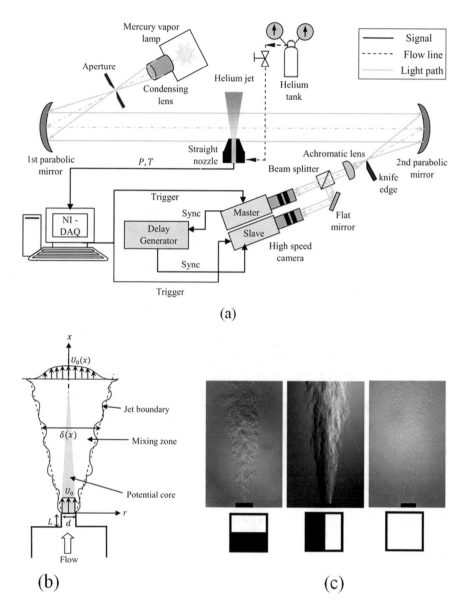

Fig. 3.1 Schematic of (**a**) an experimental setup for high-speed schlieren imaging of axisymmetric helium jet, (**b**) a free jet issuing from a straight nozzle and its coordinate system, (**c**) schlieren photographs of turbulent helium jet using a horizontal knife-edge

as shown in Fig. 3.1a was used for shadowgraph with minor modifications. No knife-edge was used as shadowgraph does not require any cutoff of refracted light.

The jet velocity at the nozzle exit, U_0, was within the range of 300–600 m/s ($0.3 <$ Mach < 0.6), and the length scale of the turbulent structures within the jet was typically in the order of millimeters. To capture such flow details and to extract a meaningful correlation between consecutive images, an extremely high framing rate was required. For example, to observe a 10×10 cm^2 ROI in a flow field flowing at a velocity of 500 m/s, it requires a minimum of 80 kHz frame rate for an optimum 10–12 pixels shift between consecutive images with a 1000×1000 pixel camera resolution. Moreover, to reduce integration effect, camera exposure time should be less than frame-to-frame time delay, Δt. To solve this problem, two identical black and white Phantom high-speed digital cameras with an identical lens system (105 mm Nikon f/2.8G macro lens) were used to capture schlieren visualization, with a time delay, Δt, in between the two cameras (master and slave). This time delay, Δt, was precisely controlled by the delay generator DG535 and could go as small as 0.1 μs. For $Re_d = 11{,}000$ and $Re_d = 22{,}000$, the time delay, Δt, between the two cameras were set 0.4 μs and 0.2 μs, respectively. Both cameras were synchronized with a master trigger to start the recording. Each frame of the slave camera was synchronized with the corresponding frame of the master camera by the time delay, Δt. Schlieren images were captured with a resolution of 1024×800 pixels with a framing rate up to 10,000 frames per second. As for the spatial resolution, the size of the evaluated area was 84×64 mm^2. Using two high-speed cameras side by side to capture schlieren images was an effective way to form a double-exposure high-speed camera. During image processing, each frame from the master camera was correlated with the corresponding frame from slave camera. Thus, corresponding image pairs from master, and slave cameras were correlated. So, the user-defined time delay, Δt, was the delay between these two images. The framing rate for both cameras was identical, and it merely dictated the number of image pairs acquired over a period.

Figure 3.1c describes the orientation of the knife-edge for schlieren. For both the knife-edge orientations, 40% cutoff of refracted light was used since 40% was the optimum cutoff that provided the best SNR. The contrast of the schlieren images would have increased with the higher percentage of light cutoff, but that would have had also decreased "particle" image size. This is discussed in detail in the data preprocessing section. In Fig. 3.1c the helium jet on the left represents horizontal knife-edge. It detected changes only in the vertical component of the refractive index, $\partial n/\partial x$, of the jet. The middle image of the helium jet shows the result using a vertical knife-edge where only changes in the horizontal component of the refraction index, $\partial n/\partial r$, were detected. Shadowgraph, the right most image of Fig. 3.1c, did not require knife-edge.

It was essential to characterize the shadowgraph and schlieren system. Shadowgraph intensity is defined as the ratio g/h, where g is the focus offset between schlieren and the focal plane and h is the distance between the collimating lens/mirror and the focal plane [21]. A g/h value of 0.31 was used for shadowgraphy. The schlieren sensitivity or the contrast sensitivity can be written as the rate of change of image contrast with respect to refraction angle, $\mathbb{S} = d\mathbb{C}/d\epsilon$. Contrast, $\mathbb{C}$, in the schlieren images can be expressed as the ratio of differential luminance, ΔE, to the general background level, E. $\mathbb{C} = \Delta E/E = \Delta a/a$, where a is the unobstructed height of the source image in the cutoff plane and Δa represents the change in a due to refraction. The refraction angle is ϵ. The schlieren sensitivity can be further simplified to $\mathbb{S} = f_2/a$, where f_2 is the focal length of the second parabolic mirror. This simple yet sufficient geometric-optical relation for sensitivity adequately characterizes the schlieren system. For both schlieren settings (horizontal knife-edge and vertical knife-edge), $f_2 \cong 1.2$ m and $a \cong 500$ μm produced a schlieren sensitivity, $\mathbb{S}$ of 2400.

3.2.2 PIV Imaging

To validate the flow field results from SIV techniques, traditional 2D PIV was carried out on this axisymmetric helium jet under identical experimental conditions. A double oscillator 532 nm Nd:YAG laser was used, which delivers an energy of 50 mJ per pulse with a pulse duration of 6 ns and at a sampling frequency of 15 Hz. The laser beam was converted into a laser sheet of 1 mm thickness. The jet was seeded with titanium dioxide (TiO_2) tracer particles with a nominal diameter, D_P of 0.4 μm. Particle Stokes number, S_k, was 0.035 and 0.07 for $Re_d = 11,000$ and $Re_d = 22,000$, respectively. A digital CCD camera from TSI Inc. (4MP-HS) with a 105 mm Nikon f/2.8G macro lens was used to capture a field of view (FOV) of approximately 84.5×64.5 mm^2. Multiple sets of 500 images were taken for each Reynolds number. PIV data acquisition was made using the TSI Inc. software. Since PIV and SIV were executed on identical experimental settings, the same PIV camera could have been used for schlieren measurements as well. However, the intention of SIV was to get a highly time-resolved measurement of the steady helium jet (and ultimately for other unsteady, transient jets), which is why two high-speed cameras were used for schlieren and shadowgraph measurements. Two high-speed cameras side by side, each at 10,000 frames per second, effectively created a double-exposure 10 kHz system. This was not possible with the PIV camera which was a double-exposure 32 fps maximum with pixel resolution of 2048×2048 pixels, even after sacrificing some of its spatial resolutions.

3.2.3 Calibration

The standard PIV calibration approach was applied to SIV as well. A planar target with a regularly spaced (0.5 mm) grid of markers was placed on the light path at the position of the schlieren object (the helium jet). It was moved by a specified distance in the out-of-plane direction to two or more positions to ensure alignment and to minimize any distortion in the two-camera system. At each position, a calibration polynomial mapping function with sufficient degrees of freedom mapped the global x-r plane to camera planes.

3.3 Data Analysis Techniques

Data analysis procedures are comprised of three major segments. Data preprocessing techniques including image filtering, inverse Abel transform, and particle size limitations are discussed in first part, followed by data processing parameters such as correlation algorithm, grid size, window size, overlapping parameter, etc. The final section discusses post-processing methods.

3.3.1 Data Preprocessing Parameters

3.3.1.1 "Particle" Image Size and "Particle" Density

In PIV, two critical parameters that control bias and root-mean-square (RMS) error are particle image size, d_p, the size of the particle image in pixels and particle density, n_p, the number of particle images per interrogation window. A particle image size less than 1 pixel creates peak locking, and higher than 3 pixels brings bias error [22]. Likewise, a lower particle density, $n_p < 8$, incurs mean bias and root-mean-square (RMS) error. To minimize errors in PIV, researchers have found a particle image size in the range of $2 < d_p < 3$ *pixels* and a particle density in the range of $8 < n_p < 22$ work best for cross-correlation-based PIV processing algorithms [23]. In PIV experiment, this particle image size is meticulously controlled by the projected pixel resolution factor (μm/pixels) during the calibration process.

An edge detection method was implemented in SIV to ensure the minimum "particle" density requirement. Marr and Hildreth [24] edge detection algorithm that combines Gaussian filtering with the Laplacian was used to detect "particle" edges. A typical SIV "particles" map is shown in Fig. 3.2 using Marr and Hildreth

Fig. 3.2 Typical SIV "particles" using Marr and Hildreth edge detection method

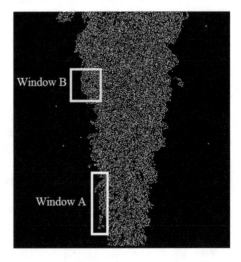

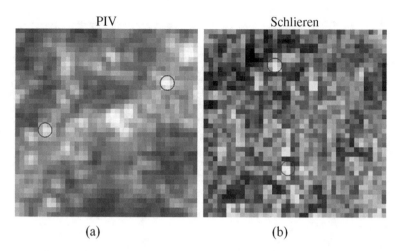

Fig. 3.3 Unfiltered 40 × 40 pixels image segments from (**a**) PIV and (**b**) schlieren showing particle image size, d_p (approximately marked by a red circle). The "particle" image size lies between $2 < d_\mathrm{p} < 3$ pixels

edge detection method. Two different interrogation windows, "window A" (rectangular) and "window B" (square), are shown on this "particles" map. A quick overview to generate a "particles" map using Marr and Hildreth algorithm is outlined here. First, a Gaussian filter was applied to the entire image. Then, zero-crossings were detected in the filtered image to obtain the edges. Zero-crossing occurs when the brightness changes over a threshold gray-scale value. This threshold gray value is user-defined, and we used the average gray value of the respective images. Marr and Hildreth edge detection method is particularly suitable when there are substantial and rapid variations in image brightness. Although Canny edge detection method [25] generally does a better job in detecting edges, the edge contours can be substantially fragmented. On the contrary, Marr and Hildreth edge detection algorithm always form connected, closed contours that help easy detection of SIV "particles."

However, in SIV techniques, there are no real particles. Rather turbulent eddies are considered as seeding "particles." Since we were using an open source processing code designed for PIV, we chose to follow these error minimizing guidelines for SIV techniques as well. Figure 3.3 shows the "particle" image size, d_p, for schlieren and PIV, respectively. For both cases, the intensity weighted-"particle" image size lied within the range of $2.6 < d_\mathrm{p} < 2.9$ pixels. However, the eddy size changes with the change in integral length scale along the axial direction of the jet. Figure 3.4 shows the variation of "particle" image sizes, d_p, along the jet in SIV. "Particle" size increases in a monotonic fashion from $d_\mathrm{p} = 1.7$ pixels at the jet exit to $d_\mathrm{p} = 3.2$ pixels at $x/d = 20$. Nevertheless, "particle" sizes fall in the range of $2 < d_\mathrm{p} < 3$ *pixels* for most of the near and intermediate regions, $4 < x/d < 16$. This was achieved by precisely controlling the projected pixel resolution factor in SIV

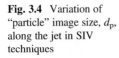

Fig. 3.4 Variation of "particle" image size, d_p, along the jet in SIV techniques

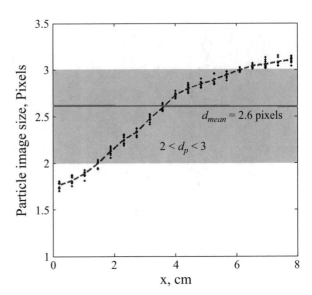

measurements. A projected pixel resolution of 80 µm/pixel was used for the current experiment.

"Particle" density, n_p, dictates the window sizes during post-processing, for both PIV and SIV techniques. Window sizes for multiple passes were chosen carefully to reduce errors. The first pass was dictated by the maximum velocity of the flow field. In other words, the largest window size was required in the first pass to resolve the highest velocity. This window size automatically included enough "particles." However, for the successive passes, the window size was reduced in a manner so that the interrogation window held just enough "particles," n_p to reduce the bias error. Window sizes of 32×128 pixels and 16×16 pixels were selected for the first pass and the final pass, respectively.

3.3.1.2   Inverse Abel Transformation

Contrary to 2D PIV where the probe volume is made of a laser sheet with a thickness of one millimeter or less, schlieren and shadowgraph produce path-integrated volumetric signals. Thus, a reverse Abel transforms [26] were necessary. For an axisymmetric jet, the local intensity i can be evaluated using the inverse Abel transformation of the line of sight intensity, I, which is obtained from the intensity measurement as a result of a change in refractive index from schlieren images. The relationship between the local intensity, i, and the line of sight intensity, I, under axisymmetric assumption can be expressed as

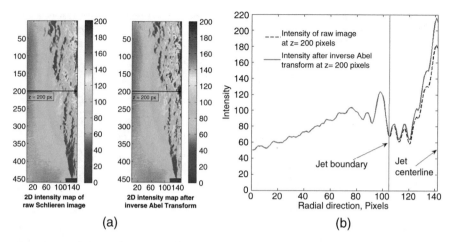

Fig. 3.5 The effect of Abel inversion on (**a**) a typical schlieren image of the turbulent round jet and (**b**) the line intensity at an axial location of $z = 200$ pixels

$$i = -\frac{1}{\pi} \int_r^\infty \frac{dI}{dy} \frac{dy}{\sqrt{y^2 - r^2}} \tag{3.1}$$

where r is the radial distance as shown in Fig. 3.1b and y is the distance from the jet centerline perpendicular to the line of sight direction. Then the local intensity i can be calculated using Eq. (3.1). Figure 3.5a shows the Abel inversion of a typical schlieren image of the turbulent round jet. Figure 3.5b compares the line intensity at an axial location $z = 200$ pixels before and after the Abel inversion.

Abel transformation requires an axial symmetry of the velocity field. The temporally and spatially averaged mean flow field fulfilled this assumption. However, the instantaneous flow field extracted from the consecutive schlieren images did not fulfill this assumption of symmetry. Thus, to apply Abel inversion on instantaneous schlieren images, it was essential to study the behavior of time averaged Abel inversion under varying sample sizes, i.e., the number of SIV images. As shown in Fig. 3.5 that for a 2D intensity map of schlieren images, Abel inversion affected the intensity values of the jet. However, at times numerical error could propagate while solving the inverse Abel transform (Eq. 3.1). Thus, an ensemble time averaging of a set of 5, 10, and 50 schlieren images for each test condition was compared to examine the noise in the Abel inversion and is shown in Fig. 3.6. We studied the variation of the Abel transformed quantity (image intensity) for three different sample sizes, $N = 5$, 10, and 50. As we increased the number of images to $N = 50$, the inversion attained a mean value. However, as we decreased the sample size to $N = 5$, nominal scatter was observed around the mean, but overall the variation was relatively small as evident from Fig. 3.6.

Since the Abel inversion only negates the effect of volumetric path integration nature of schlieren images, once we achieved a steady inversion, unless we changed the experimental conditions such as light intensity, knife-edge cutoff, and camera

Fig. 3.6 Abel inversion of schlieren intensity counts for different sample sizes

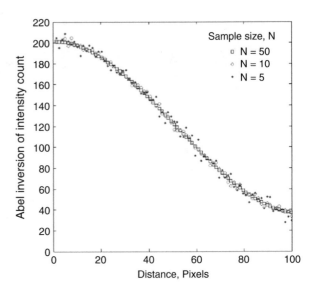

settings, the variation in Abel transformed quantity (image intensity) was negligible. Previous studies [27–30] showed the potential of using Abel inversion to transform an instantaneous axisymmetric measurement into the planar measurement.

3.3.1.3   Image Filtering

Image restoration and enhancement play a significant role in both traditional PIV as well as for SIV techniques. Traditional PIV has proven preconditioning techniques to attenuate or completely remove undesired effects via image reconstruction (e.g., background subtraction) or image enhancement (e.g., min-max contrast normalization). Since the physical processes behind schlieren and shadowgraph techniques are completely different than PIV, not all image reconditioning methods commonly used for PIV can be implemented on schlieren images. A variety of image filtering techniques were tested, and some of them are discussed in the following.

The quality of the present PIV images suffered due to pulse-to-pulse variation of the light intensity from the Nd:YAG laser. Intensity normalization fixed this issue. PIV images also contained dark current and thermal noise of a CC-D sensor. Due to homogeneous statistical properties of this noise, a background subtraction greatly reduced it. Background subtraction also reduced the effects of laser flare and other stationary image features. Lastly, to smooth out small-scale intensity fluctuations, a uniform filter, which is a linear low-pass filter, was used to replace each pixel by the average gray value over a 3×3 pixel subdomain. After applying the uniform filter, the resulting image was subtracted from the original image. Resulted improvements are shown in Fig. 3.7a along with the gray value histograms for a PIV image segment. The left image of Fig. 3.7a shows a noisy PIV image segment with random

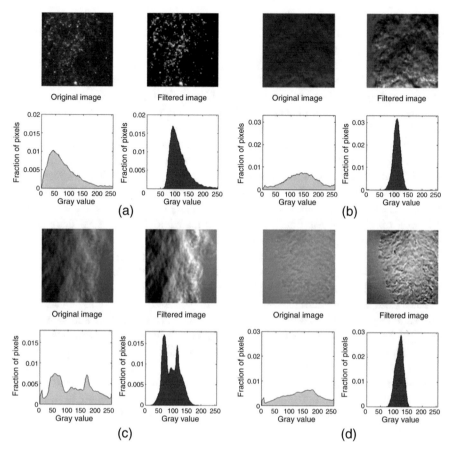

Fig. 3.7 Improvement of image segmentation and gray value histograms for (**a**) PIV, (**b**) schlieren with horizontal knife-edge, (**c**) schlieren with vertical knife-edge, and (**d**) shadowgraph

noise fluctuations as evident from the widely distributed gray value histogram. After an intensity normalization, background subtraction, and uniform filtering, the gray value histogram reduced in width, as shown in the right side of Fig. 3.7a. Similar image improvements were obtained using a Gaussian filter instead of the uniform filter as well.

PIV image intensity field involved bright particle images on a darker background, whereas the schlieren and shadowgraph images contained turbulent structures on a gray color background. It was impossible to use background subtraction on schlieren images without removing some of the actual signals. That is why image enhancement or amplification served best rather than image reconstruction for SIV techniques.

For all SIV images, first, a narrow-width, low-pass filter was applied to remove high-frequency noises (e.g., camera shot noise, pixel anomalies, and digitization artifacts) from the images. This also allowed the subpixel peak fitting algorithm to perform better by widening the correlation peaks. Contrast enhancement increased signal content in SIV images and helped to find the definite correlation between successive images. In other words, optimization of image contrast was necessary for SIV. A nonlinear min-max filter approach as suggested by Westerweel [31] was used to normalize the image contrast. Lastly, to enhance the edges of the turbulent structures, Laplace filter [32] was applied. Due to the highest velocity near the jet tip, there was always a chance that a longer camera exposure time could blur turbulent structures and could create an under-sampled image segment. This could lead to the so-called peak-locking effect. However, a high-pass Laplace filter reduced this adverse effect. The combined effects of all these filters are shown in Fig. 3.7b–d for schlieren with a horizontal knife-edge, schlieren with vertical knife-edge, and shadowgraph, respectively. Gray value histograms behave fairly similarly for schlieren with horizontal cutoff and shadowgraph as gray levels narrow down between 60 and 150 for filtered images compared to their original counterparts spreading across entire 0–255 gray value range.

While the gray value histograms of filtered image segments for schlieren with horizontal knife-edge and shadowgraph demonstrated improvements over the original image, schlieren with vertical knife-edge failed to do so. Vertical knife-edge produced an uneven illumination on left and right side of the schlieren image as shown in Fig. 3.1c. Thus, double-peaked gray value histogram appeared as depicted in Fig. 3.7c. Even with this limitation, vertical knife-edge schlieren was investigated in this study for the comparison purposes between different SIV techniques.

For white-light schlieren, the darkness of the gray background increased with increase in knife-edge cutoff. An increasing cutoff also enhanced image contrast. However, this increase in the contrast came at the cost of lower "particle" size with higher cutoff [14]. This was the prime reason why 40% cutoff for both the horizontal and vertical knife-edge schlieren was chosen in the first place.

3.3.2   Data Processing Parameters

The data was post-processed using an open source GUI-driven Matlab code, quantitative imaging (QI), or Prana [33] originally developed at the AEThER Laboratory of Virginia Tech. Robust phase correlation (RPC) [19] algorithm was used, which essentially uses the phase of the Fourier transform-based cross-correlation plane and applies a Gaussian spectral filter to optimize the SNR. The use of RPC algorithm becomes justified from Fig. 3.3b that showed the intensity distribution of turbulent structures from schlieren images. Intensity map of turbulent eddies in schlieren images looked approximately Gaussian. A higher-order multigrid CWO (continuous window offset) iterative image deformation method was applied throughout, for which the image window was deformed to diminish the loss of information due to shear and/or rotation of the image windows. Additionally, bi-cubic interpolation for velocity field interpolation and the cardinal function with Blackman filter for image interpolation were employed. To obtain a converged velocity field, a total of four to five iterations with varying processing parameters such as grid size, window size, resolution, and vector validation was used. To resolve higher velocities at the jet exit, a rectangular window of 1:4 ratio (32×128 pixels, 16×64 pixels) where the long side of the window aligned streamwise (or in the x direction) was used on the first pass. This generated 50% overlap in the radial direction and 87.5% overlap in the streamwise direction. A 1:1 16×16 pixels interrogation window size with 50% overlap at the final pass was applied, leading to a vector spacing of 0.4 mm for the time-resolved measurements. Grid resolution of 8×8 pixels and a particle image diameter of 2.8 pixels were used. The three-point Gaussian estimator was employed for the subpixel correlation peak location for interpolating the correlation peak location (and hence displacement) below the specified pixel resolution.

3.3.3 Data Post-processing Parameters

3.3.3.1 Validation of Velocity Field

Post-processing parameters were set up separately for each pass for all measurement techniques. Two essential validation techniques, velocity thresholding and universal outlier detection (UOD), were implemented to obtain the final velocity field for all passes except the final pass. To remove high-frequency noise signals, a Gaussian smoothing filter was applied to each vector field by taking an average value of the neighborhood grid points.

3.4 Results and Discussion

The velocity fields of the helium jet obtained from PIV and SIV techniques are discussed in Sect. 3.4.1. Section 3.4.2 compares the spatial variation of correlation planes from different SIV methods with PIV. Lastly, a statistical survey of different methods is conducted in Sect. 3.4.3.

3.4.1 Helium Jet Results

An Abel inversion was applied during SIV data processing to find the true velocity field. Figure 3.8a shows that an Abel inversion was necessary for the SIV to yield comparable results to the PIV data. The effect of Abel inversion is maximum at the jet centerline, $r = 0$, since the effect of path integration reaches highest at the centerline. The Abel inversion was applied on schlieren images at the beginning of the data processing, even before applying any image enhancement/restoration technique. Several past studies [14, 34] performed this step, an inverse Abel transform,

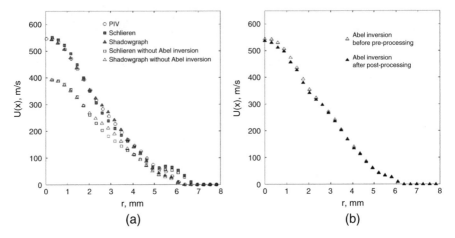

Fig. 3.8 (a) Comparison of SIV results with PIV before and after Abel inversion and (b) Abel inversion at different stages of data processing for $x/d = 10$ of $Re_d = 22{,}000$ jet

during the post-processing of SIV data. Thus, we wanted to understand and validate the effect of Abel inversion at two different stages of data processing, (1) at the beginning of the preprocessing, before applying image filters, and (2) after the processing (cross-correlation) and data validation, at the very last step. Figure 3.8b compares Abel inversion of a schlieren velocity profile of $Re_d = 22{,}000$ jet at two different stages of data processing. It shows both velocity profiles agreed well. However, the Abel inversion before image preprocessing showed a slightly better agreement with PIV measurements.

Figure 3.9 compares the instantaneous and mean velocity fields of the helium jet of $Re_d = 11{,}000$ and $Re_d = 22{,}000$ obtained from PIV and various SIV methods. The sizes of the velocity vectors are proportional to their corresponding magnitudes, and the vectors are plotted on a velocity magnitude contour map. For every measurement technique, 500 image pairs had been used to calculate the mean flow field. Schlieren with horizontal knife-edge and shadowgraph showed excellent agreement with the PIV results. However, schlieren with vertical knife-edge failed to provide a meaningful velocity field. As mentioned earlier, the schlieren with vertical knife-edge detects only the horizontal components, $\partial n/\partial r$, of the refractive index. This produced asymmetric illumination in the transverse direction of the jet. Due to asymmetric illumination, the "particle" size increases from the dark to the bright region in vertical knife-edge schlieren. The larger size "particles" incur bias error in the calculated velocity field. Even though several image filters were tested, they failed to restore any useable signals from vertical knife-edge schlieren images. A detailed analysis of the poor performance of vertical knife-edge schlieren is presented using the correlation plane statistics in subsequent sections. For a better quantitative comparison, velocity magnitude isolines were plotted on the mean flow contour for $Re_d = 22{,}000$ as shown in Fig. 3.10.

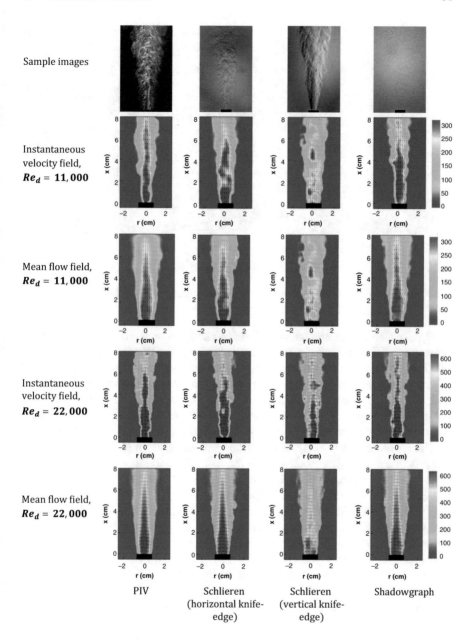

Fig. 3.9 Instantaneous and mean flow field colored by velocity magnitude from PIV, schlieren with horizontal knife-edge, schlieren with vertical knife-edge, and shadowgraph for two different flow conditions, $Re_d = 11,000$ and $Re_d = 22,000$

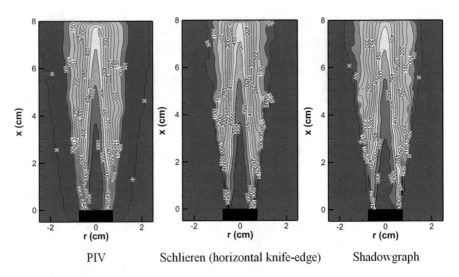

PIV   Schlieren (horizontal knife-edge)   Shadowgraph

Fig. 3.10 Isolines of velocity magnitude on the mean flow field for $Re_d = 22,000$

PIV, schlieren with horizontal knife-edge and shadowgraph, all yielded comparable results. Time average jet centerline velocities at the nozzle exit were 611 m/s,
609 m/s, and 608 m/s for schlieren with a horizontal knife-edge, shadowgraph, and
PIV, respectively. The length of the potential core was approximately 3.7 cm for
both PIV and horizontal knife-edge schlieren. Shadowgraph, however, showed a
potential core length of 3 cm, shorter than PIV or schlieren. But all three measurement techniques agreed very well at the downstream of the jet.

The jet centerline velocity at the nozzle exit were 304 m/s (Mach = 0.3) and
611 m/s (Mach = 0.6), respectively, for $Re_d = 11,000$ and $Re_d = 22,000$. The mean
axial velocity, U, normalized by the centerline jet velocity, U_0, is plotted in
Fig. 3.11a, b for both the Reynolds numbers.

The near-field velocity profiles obtained from PIV, horizontal knife-edge schlieren, and shadowgraph were compared to flying and stationary hot-wire measurements by Fellouah [35] for the Reynolds number ranging from 6000 to 30,000 on the
near and intermediate region ($0 \le x/d \le 25$) of a round free jet. The comparison
showed that the centerline velocity data of the turbulent helium jet obtained by
horizontal knife-edge schlieren and shadowgraph agreed very well with the PIV
measurements as well as with Fellouah [35]. The axial jet velocity profiles appeared
symmetric within experimental uncertainty. The jet tip velocity profile closely
resembled the plug flow assumption. The velocity distribution retained a top-hat
shape at $1.5 - 3d$ downstream of the nozzle exit for $Re_d = 11,000$ and $Re_d = 22,000$.
The axial mean velocity profile at $x/d = 5$, $Re_d = 11,000$ was not as developed as
seen for $Re_d = 22,000$, as it retained remnants of the initial conditions. This
observation is consistent with the widely known fact that the length of the potential
core decreases with increasing Reynolds number [36].

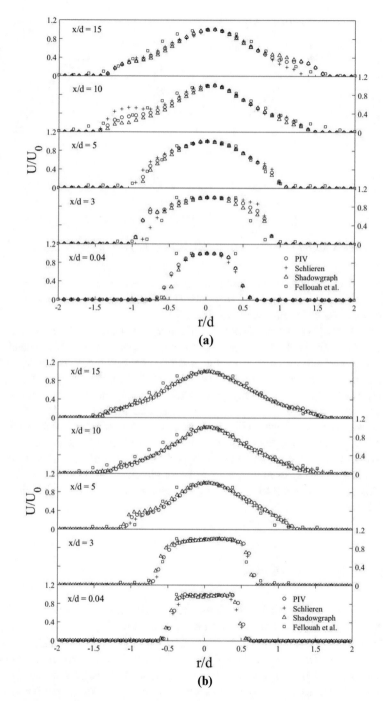

Fig. 3.11 Normalized streamwise mean velocity, U/U_0, at different axial positions for (**a**) $Re_d = 11,000$ and (**b**) $Re_d = 22,000$

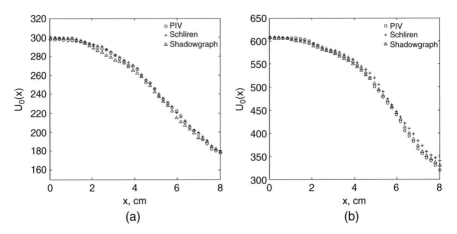

Fig. 3.12 Mean centerline velocity decay for (**a**) $Re_d = 11{,}000$ and (**b**) $Re_d = 22{,}000$

Figure 3.12 compares the mean centerline velocity decay of PIV and SIV techniques for $Re_d = 11{,}000$ and $Re_d = 22{,}000$. Schlieren with a horizontal knife-edge and shadowgraph agreed well with PIV.

Figure 3.13a, b show the streamwise turbulence intensity, u'/U_0, at several axial locations for both the Reynolds numbers using various PIV and SIV techniques. Turbulence intensity was less inside the potential core and increased toward the jet boundary near the developing region. As the jet developed, the centerline turbulence intensity started growing. At any given axial location, the centerline turbulence intensity was lower for $Re_d = 22{,}000$ as compared to $Re_d = 11{,}000$. Reynolds shear stress, $<u'v'>$, is shown in Fig. 3.14, normalized by the square of the centerline velocity, U_0^2. The highest percentage of momentum exchange, which involves large-scale vortices, occurred in the jet shear layer. The location of maximum Reynolds stress got shifted away from the jet centerline with increased downstream distance. Reynolds stress calculated from SIV techniques agreed well with PIV measurements.

3.4.2 Correlation Planes

Figure 3.15 shows the spatial correlations of the jet with $Re_d = 22{,}000$ obtained from PIV and SIV techniques at $x/d = 10$ for three radial locations $\frac{r}{d} = 0, 0.5, \ 1, and \ 1.5$, respectively. It is evident from the Fig. 3.15 that PIV showed excellent spatial correlation at the near-field region of the jet ($r/d = 0, 0.5$), but the spatial correlations became noisy at $r/d = 1$ and 1.5. For schlieren with horizontal knife-edge and shadowgraph, the spatial correlations showed similar behavior. These correlations appeared noisier than PIV. However, the presence of a single clear peak was prominent for all the radial locations. Spatial correlations from all the SIV techniques

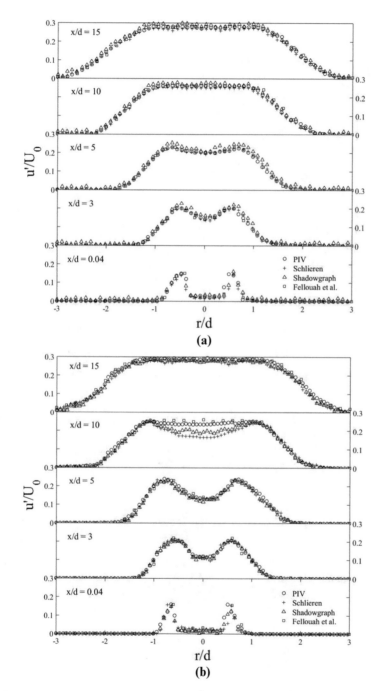

Fig. 3.13 Streamwise turbulence intensity, u'/U_0, at different axial locations for (**a**) $Re_d = 11{,}000$ and (**b**) $Re_d = 22{,}000$

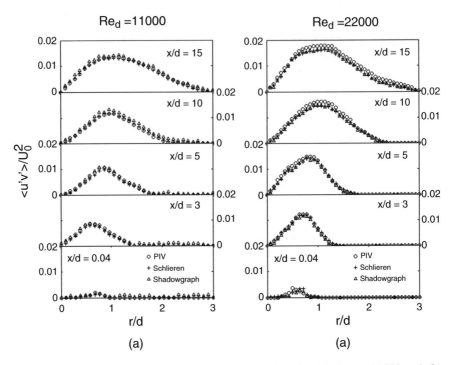

Fig. 3.14 Reynolds shear stress at different axial locations for (**a**) $Re_d = 11,000$ and (**b**) $Re_d = 22,000$

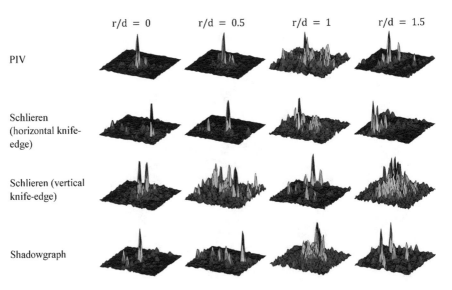

Fig. 3.15 Comparison of the spatial correlations of $Re_d = 22,000$ jet from PIV, schlieren with horizontal knife-edge, schlieren with vertical knife-edge, and shadowgraph at $x/d = 10$ for various radial locations

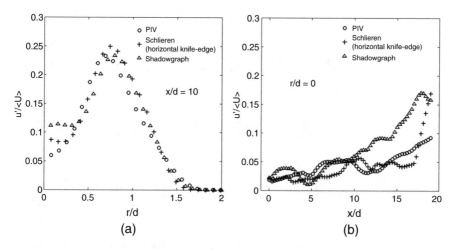

Fig. 3.16 Streamwise turbulence intensity, $u'/\langle U \rangle$ variation along (**a**) radial direction for $x/d = 10$ and (**b**) along axial direction for $r/d = 0$ of $Re_d = 22{,}000$ jet

exhibited increasing noise in the correlation information at $r/d = 1$. This observation can be explained by looking at the turbulence intensity at $r/d = 1$.

Figure 3.16a plots the variation of the streamwise turbulence intensity, $u'/\langle U \rangle$, along the radial direction at $x/d = 10$, where u' is the streamwise velocity fluctuation and $\langle U \rangle$ is the mean velocity at that location. Since the contribution of the transverse turbulence intensity, $v'/\langle U \rangle$, to the overall turbulence intensity at any given location was an order of magnitude smaller [37], only the effect of streamwise turbulence intensity was plotted and discussed here. In the developing region of the jet $x/d = 10$, the streamwise turbulence intensity maximized within the region of $0.7 \leq r/d \leq 0.8$. Thus, the spatial correlations near $r/d = 1$ became noisy due to the presence of strong turbulence. Turbulence intensity dropped quickly after $r/d = 1.3$ and reached 10% of its maximum value at $r/d = 1.42$. However, this did not explain the correlation noises at $r/d = 1.5$. The noises at $r/d = 1.5$ were primarily due to lower seeding density. Perhaps a sophisticated scheme such as adaptive windowing and multiframe was required to adapt this high-gradient region to reduce noise. Bias error and RMS error increased with a lesser number of "particles." As the number of "particles" decreased toward the jet edge (or in the mixing layer), spatial correlations got noisy. This held true for both horizontal knife-edge schlieren and shadowgraph.

Figure 3.17 shows spatial correlations for the same experimental conditions and image velocimetry techniques at the jet centerline ($r/d = 0$) for four different streamwise locations, $x/d = 0.04$, 5, 10, and 20, respectively. Figure 3.17 exhibits the presence of clear, unambiguous peaks in the correlation planes near the jet centerline ($r/d = 0$) for PIV, horizontal knife-edge schlieren, and shadowgraph. Spatial correlations behaved in excellent fashion inside the potential core, $0.04 \leq x/d \leq 10$. However, SNR gradually went down for $x/d > 10$. This happened because as the jet entered the developing region from the potential core, the streamwise turbulence intensity increased. Figure 3.16b shows the streamwise turbulence

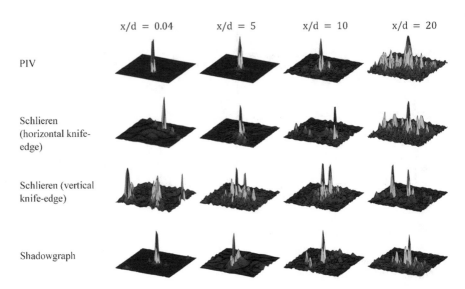

Fig. 3.17 Comparison of the spatial correlations of $Re_d = 22{,}000$ jet from PIV, schlieren with horizontal knife-edge, schlieren with vertical knife-edge, and shadowgraph at jet centerline ($r/d = 0$) for various axial location

intensity variation along the jet centerline. The turbulence intensity increased nearly linearly with x/d. Turbulence intensity increased slowly between $0.04 \leq \frac{x}{d} \leq 10$, and it grew rapidly afterward. An increase in turbulence intensity for $x/d > 10$ induced noise in the spatial correlations.

As seen in the earlier sections, schlieren with vertical knife-edge failed to provide accurate velocity information. Correlation statistics discussed above could explain it from a statistical standpoint. Figures 3.15 and 3.17 show schlieren with vertical knife-edge had ambiguity in displacement correlation, the presence of multiple peaks of the same order, lower peak-to-peak ratio, higher background noise, and lower SNR.

3.4.3   PPR, PCE, and SNR PDF

Cross-correlation SNR can be expressed as the ratio between the primary (the tallest) peak, $|\mathbb{C}_{max}|$, and the second (next tallest) peak, $|\mathbb{C}_2|$, in the correlation plane. This can be written as

$$\text{PPR} = \frac{|\mathbb{C}_{max}|}{|\mathbb{C}_2|} \tag{3.2}$$

This ratio is termed as the primary peak ratio (PPR) and is used as a measure of the detectability of the true displacement [38, 39]. Often a user-defined threshold

Fig. 3.18 Separation of cross-correlation plane from background noise. (**a**) Cross-correlation plane with background noise, (**b**) the correlation plane related to background noise, (**c**) cross-correlation plane without background noise

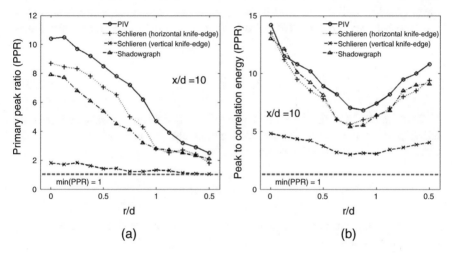

Fig. 3.19 Radial variation of (**a**) the primary peak ratio (PPR), the ratio between the primary correlation peak to the second tallest peak, (**b**) the peak to correlation energy (PCE) for PIV, schlieren with horizontal knife-edge, schlieren with vertical knife-edge, and shadowgraph at $x/d = 10$ of $Re_d = 22,000$ jet

(such as 1.2 or 1.3) for PPR is considered to validate the correlation. To remove the effect of any unwanted background noise in calculating PPR, the correlation planes were separated from the background noise. A typical cross-correlation plane separated from its background noise is represented in Fig. 3.18. The corresponding cross-correlation planes of the two image sets, with and without background noise, were subtracted one from the other to obtain the background noise.

PPR for all image velocimetry techniques was compared along the radial direction at $x/d = 10$. PPR was found to be high, between 8.2 and 10.5 near the centerline at $r/d = 0$ for all measurement techniques except schlieren with a vertical knife-edge, as shown in Fig. 3.19a. Due to lower turbulent fluctuations at the jet centerline, spatial correlations detected unambiguous displacements of the turbulent structures or tracer "particles"; hence PPR increased near the centerline. However, as we walked away from the centerline, PPR decreased almost linearly and reached a value of PPR $\approx$ 2 near the jet boundary, $r/d = 1.5$. This is in good agreement with Hain [40] who suggested that a threshold PPR, PPR $\geq$ 2, can reliably avoid the spurious vectors. Figure 3.19a shows that the highest PPR was found in PIV, although schlieren with horizontal knife-edge and shadowgraph showed superior quality PPR as well. Another important observation was that PPR in shadowgraph was lower than horizontal knife-edge schlieren for $0 \leq r/d < 1$. These observations can be attributed to the fact that schlieren produced higher contrast images, which in turn helped to obtain best correlations. However, after $r/d \geq 1$ PPR for both of these measurement techniques attained similar values.

PPR is an ad hoc representation for SNR and the correlation plane statistics. It does not include information of the entire correlation plane. A better way to quantify the correlation SNR is to use the peak signal to correlation energy (PCE) defined as the ratio between the magnitude of the cross-correlation plane and the correlation energy as

$$PCE = \frac{|\mathbb{C}_{max}|^2}{E_c} \tag{3.3}$$

The magnitude of the correlation plane $|\mathbb{C}_{max}|$ is the signal part representing the tallest peak in the correlation plane. Alternatively, the correlation energy which defines the noise part can be expressed as

$$E_c = \int_{-\infty}^{\infty} |\mathbb{C}_{max}|^2 dx \tag{3.4}$$

However, the correlation plane has a finite size. Thus, the correlation energy of a finite size correlation plane can be calculated as

$$E_c = \frac{1}{W}\left(\sum_W |\mathbb{C}_{max}|^2\right) \tag{3.5}$$

where W is the size of the correlation plane. Figure 3.19b plots the radial variation of PCE for the similar settings as used to find PPR. PCE shows similar behavior for all measurement techniques except schlieren with a vertical knife-edge. PCE was found higher near the jet centerline and at the boundary/mixing layer, but PCE reached a minimum for $0.6 < r/d < 1$. This occurred as the streamwise turbulence intensity affected PCE in a similar fashion as PPR, as shown in Fig. 3.16a. The streamwise turbulence intensity increased by 20–25% between $0.6 < r/d < 1$, which affected correlation plane statistics and in turn peak delectability.

To identify the correct velocity vector, the amplitude of the displacement correlation peak (signal) must be larger than the amplitude of the tallest random peak (noise), SNR $\gg$ 1. To evaluate the probability of SNR $\gg$ 1, normalized probability density functions (PDF) of signal and noise peak amplitudes from the entire

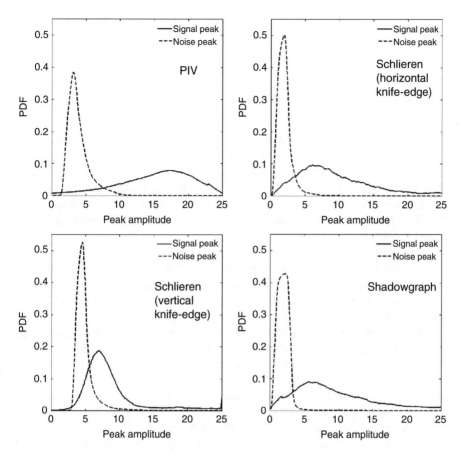

Fig. 3.20 The probability density functions (PDF) for the normalized amplitude of the displacement correlation peak (signal) and the tallest random correlation peak (noise) for PIV, schlieren with a horizontal knife-edge, schlieren with vertical knife-edge, and shadowgraph

correlation plane were plotted in Fig. 3.20 for PIV and SIV techniques. The peak amplitude of the signal was much wider than the noise. Also, the mean signal peak amplitude shifted significantly for different measurement techniques. Both signal and noise peak amplitude PDFs for schlieren with horizontal knife-edge and shadowgraph showed similar behavior. Signal peak amplitude shifting to far right for PIV indicated higher signal density compared to SIV techniques. Like the signal, noise peak amplitude also increased for PIV and shifted toward the right and became wider, as compared to SIV techniques. However, this increase in noise level was insignificant relative to the increase in signal intensity for PIV. A lesser overlap between signal and noise peaks signified the better possibility of correlation (or in other words, increasing the probability of identification of the tallest displacement peak). An excess overlap of signal and noise peaks for schlieren with vertical knife-edge was responsible for producing spurious displacement peaks during spatial correlation as shown in Figs. 3.15 and 3.17.

3.5 Conclusions

Current research demonstrated a novel, inexpensive, easy to setup two-camera SIV technique that could resolve exceptionally high flow velocities. Statistical assessment of SIV techniques was performed for a high-velocity helium jet at two different Reynolds numbers, $Re_d = 11,000$ and $Re_d = 22,000$. The velocity field obtained by horizontal knife-edge schlieren with 40% cutoff and shadowgraph agreed well with the PIV results. Vertical knife-edge schlieren with 40% cutoff performed poorly due to inconsistent signal content.

Three filters were applied that improved the SNR of schlieren images, (a) a narrow-width, low-pass filter such as a uniform filter to remove high-frequency noise, (b) a nonlinear min-max filter to normalize the image contrast, and (c) a Laplace filter to enhance and sharpen the turbulent structures. A "particle" image size, d_p, between 2 and 3 pixels proved to work best.

The performance of the spatial correlations was quantitatively evaluated using PPR and PCE. A high value of PPR was observed near the jet centerline for schlieren with horizontal knife-edge and shadowgraph. PPR decreased linearly from jet centerline toward shear/mixing layer but retained PPR > 1. Due to higher turbulence intensity, the value of PCE dropped between $0.6 < r/d < 1$. PDF of signal and noise showed that PIV included higher noise than schlieren. However, the signal content in PIV was two- to threefold higher than schlieren measurements.

SIV demonstrated promising results toward seedless velocity measurements, although it has some limitations. Due to the path-integrated nature of current schlieren/shadowgraph techniques, it works best for axisymmetric or 2D flows but not for complex 3D flows. It does not work for laminar flows because of lack of turbulent structures. However, a focusing schlieren system can potentially be used to measure a three-dimensional flow field.

References

1. Papamoschou, D.: A two-spark schlieren system for very-high velocity measurement. Exp. Fluids. **7**(5), 354–356 (1989)
2. Wu, Y.: Detection of velocity distribution of a flow field using sequences of Schlieren images. Opt. Eng. **40**(8), 1661 (2001)
3. Fu, S., Wu, Y.: Quantitative analysis of velocity distribution from schlieren images. In: Carlomagno, G.M., Grant, I. (eds.) Proceedings of the 8th International Symposium on Flow Visualization, Sorrento (1998)
4. Raffel, M.: Background-Oriented Schlieren (BOS) techniques. Exp. Fluids. **56**(3), 1–17 (2015)

5. Raffel, M., Richard, H., Meier, G.E.A.: On the applicability of background oriented optical tomography for large scale aerodynamic investigations. Exp. Fluids. **28**(5), 477–481 (2000)
6. Raffel, M., et al.: Background oriented stereoscopic schlieren for full-scale helicopter vortex characterization. In: Carlomagno, G.M., Grant, I. (eds.) Proceedings of 9th International Symposium on Flow Visualization, Edinburgh (2000)
7. Elsinga, G.E., et al.: Assessment and application of quantitative schlieren methods: calibrated color schlieren and background oriented schlieren. Exp. Fluids. **36**(2), 309–325 (2003)
8. Elsinga, G.E., et al.: Assessment and application of quantitative schlieren methods with bi-directional sensitivity: CCS and BOS. In: Proceedings of PSFVIP-4, Chamonix (2003)
9. Scarano, F., Benocci, C., Riethmuller, M.L.: Pattern recognition analysis of the turbulent flow past a backward facing step. Phys. Fluids. **11**(12), 3808–3818 (1999)
10. Kindler, K., et al.: Recent developments in background oriented Schlieren methods for rotor blade tip vortex measurements. Exp. Fluids. **43**(2–3), 233–240 (2007)
11. Goldhahn, E., Seume, J.: The background oriented schlieren technique: sensitivity, accuracy, resolution and application to a three-dimensional density field. Exp. Fluids. **43**(2), 241–249 (2007)
12. Kegerise, M.A., Settles, G.S.: Schlieren image-correlation velocimetry and its application to free-convection flows. In: Carlomagno, G.M., Grant, I. (eds.) 9th International Symposium on Flow Visualization, pp. 1–13. Heriot-Watt University, Edinburgh (2000)
13. Garg, S., Settles, G.S.: Measurements of a supersonic turbulent boundary layer by focusing schlieren deflectometry. Exp. Fluids. **25**(3), 254–264 (1998)
14. Jonassen, D.R., Settles, G.S., Tronosky, M.D.: Schlieren "PIV" for turbulent flows. Opt. Lasers Eng. **44**(3–4), 190–207 (2006)
15. Hargather, M.J., Settles, G.S.: A comparison of three quantitative schlieren techniques. Opt. Lasers Eng. **50**(1), 8–17 (2012)
16. Hargather, M.J., et al.: Seedless velocimetry measurements by Schlieren Image Velocimetry. AIAA J. **49**(3), 611–620 (2011)
17. Mauger, C., et al.: Velocity measurements based on shadowgraph-like image correlations in a cavitating micro-channel flow. Int. J. Multiphase Flow. **58**, 301–312 (2014)
18. Zelenak, M., et al.: Visualisation and measurement of high-speed pulsating and continuous water jets. Measurement. **72**, 1–8 (2015)
19. Eckstein, A., Vlachos, P.P.: Digital Particle Image Velocimetry (DPIV) robust phase correlation. Meas. Sci. Technol. **20**(5), 055401 (2009)
20. Pope, S.B.: Turbulent Flows, vol. xxxiv, p. 771. Cambridge University Press, Cambridge, UK/New York (2000)
21. Settles, G.S.: Schlieren and shadowgraph techniques. In: Experimental Fluid Mechanics. Springer-Verlag, Berlin (2001)
22. Fincham, A.M., Spedding, G.R.: Low cost, high resolution DPIV for measurement of turbulent fluid flow. Exp. Fluids. **23**(6), 449–462 (1997)
23. Huang, H., Dabiri, D., Gharib, M.: On errors of digital particle image velocimetry. Meas. Sci. Technol. **8**(12), 1427–1440 (1997)
24. Marr, D., Hildreth, E.: Theory of edge detection. Proc. R. Soc. Lond. Ser. B Biol. Sci. **207**(1167), 187–217 (1980)
25. Canny, J.: A computational approach to edge detection. IEEE Trans. Pattern Anal. Mach. Intell. **8**(6), 679–698 (1986)
26. Arfken, G., Weber, H.J.: Mathematical Methods for Physicists. Academic Press, Orlando (2005)
27. Coppalle, A., Joyeux, D.: An optical technique for measuring mean and fluctuating values of particle concentrations in round jets. Exp. Fluids. **16**(3), 285–288 (1994)
28. Yildirim, B.S., Agrawal, A.K.: Full-field measurements of self-excited oscillations in momentum-dominated helium jets. Exp. Fluids. **38**(2), 161–173 (2005)
29. Joshi, A., Schreiber, W.: An experimental examination of an impulsively started incompressible turbulent jet. Exp. Fluids. **40**(1), 156–160 (2005)

30. Mayrhofer, N., Woisetschläger, J.: Frequency analysis of turbulent compressible flows by laser vibrometry. Exp. Fluids. **31**(2), 153–161 (2001)
31. Westerweel, J.: Digital Particle Image Velocimetry – Theory and Application. Delft University, Delft (1993)
32. Rosenfeld, A., Kak, A.C.: Digital picture processing. In: Computer Science and Applied Mathematics. Morgan Kaufmann, Burlington (1982)
33. Pavlos, V.A.: Qi - Quantitative Imaging (PIV and more). (2015). Available from: https://sourceforge.net/projects/qi-tools/?source=navbar
34. Watt, D.W., et al.: Theory and application of quantitative, bidirectional color schlieren for density measurement in high speed flow. Opt. Diagn. Fluids Solids Combust. **5191**(1), 145–155 (2003)
35. Fellouah, H., Ball, C.G., Pollard, A.: Reynolds number effects within the development region of a turbulent round free jet. Int. J. Heat Mass Transf. **52**(17–18), 3943–3954 (2009)
36. Weisgraber, T.H., Liepmann, D.: Turbulent structure during transition to self-similarity in a round jet. Exp. Fluids. **24**(3), 210–224 (1998)
37. Bogey, C., Bailly, C.: Large eddy simulations of transitional round jets: influence of the Reynolds number on flow development and energy dissipation. Phys. Fluids. **18**(6), 065101 (2006)
38. Keane, R.D.: Optimization of particle image velocimeters: II. Multiple pulsed systems. Meas. Sci. Technol. **2**(10), 963–974 (1991)
39. Keane, R.D.: Optimization of particle image velocimeters. I. Double pulsed systems. Meas. Sci. Technol. **1**(11), 1202–1215 (1990)
40. Hain, R., Kähler, C.J.: Fundamentals of multiframe particle image velocimetry (PIV). Exp. Fluids. **42**(4), 575–587 (2007)

Chapter 4
Supersonic Jet Ignition

Contents

4.1   Introduction

Motivated by the fact that turbulent jets from straight nozzles could ignite a lean $(0.5 < \phi < 0.9)$ main chamber reliably as discussed in Chap. 2, we wanted to explore the possibility to reach ultra-lean limit using supersonic jets. The same experimental setup that uses a dual-chamber design (a small pre-chamber resided within the big main chamber) was used except the straight nozzles were replaced by converging or converging-diverging (C-D) nozzles. The primary focus was to reveal the characteristics of supersonic jet ignition, in comparison to subsonic jet ignition. Another intention behind supersonic jets was from ignition delay standpoint; a high-speed jet could well reduce the ignition delay. Simultaneous high-speed schlieren photography and OH* chemiluminescence were applied to visualize the supersonic jet penetration and ignition processes in the main chamber. Infrared imaging was

© Springer International Publishing AG, part of Springer Nature 2018   65
S. Biswas, *Physics of Turbulent Jet Ignition*, Springer Theses,
https://doi.org/10.1007/978-3-319-76243-2_4

used to characterize the thermal field of the hot jet. Numerical simulations were carried out using the commercial CFD code, Fluent 15.0, to characterize the transient supersonic jet, including spatial and temporal distribution of species, temperature and turbulence parameters, velocity, Mach number, turbulent intensity, and so on. The present work focuses on the effect of supersonic jets on lean flammability limits.

Only a few studies have been conducted to understand the effect of high-speed, transonic, or supersonic hot reacting jets on ignition of ultra-lean fuel/air mixtures. Djebaili [1] investigated ignition of the lean H_2/air mixture by a hot supersonic (Mach $= 2.0$) jet generated in a shock tube. They observed that the flame propagation velocity inside the main chamber increased significantly using supersonic jets. Boretti [2, 3] numerically studied ignition of H_2/air using a high-speed (Mach $= 0.7$) compressible gas jet. The results showed that the ignition delay in the main chamber decreased with higher jet velocities. Chiera [4] reached similar conclusions as Boretti [2, 3] that higher velocity of the hot jet generated by pre-chamber combustion resulted in greater turbulence and multiple flame fronts, leading to faster combustion in the main chamber.

These limited studies motivated the authors to explore the concept of using supersonic hot jets to ignite ultra-lean mixtures. The authors developed an experiment [5, 6] which used a dual-chamber design to compare the ignition characteristics of subsonic jets versus supersonic jets. They found that the supersonic jets not only shortened the ignition delay but also extended the lean flammability limit of H_2/air mixture in the main chamber, as compared to the subsonic jets. For example, the lean limit achieved by subsonic jets using straight nozzles was found to be $\phi = 0.31$. Supersonic converging-diverging nozzles, however, can extend this limit to $\phi = 0.22$. Moreover, radiation intensity measurements of the hot jets showed the presence of a high-temperature zone at the downstream of the shock structures for the supersonic jets. Ignition in the main chamber was initiated near to that high-temperature zone. Our previous study [5] on the fundamental mechanism of hot jet ignition showed that both temperature and velocity of the hot jet govern the ignition process in the main chamber. Whether ignition can happen or not depends on the competition between turbulent mixing and ignition chemistry. For ignition using supersonic jets, there are additional complexities due to the presence of shock structures and the interactions between the shocks and the cold surroundings. It is essential to understand the physics behind the supersonic jet ignition process – how and why supersonic jets can extend the lean flammability limit of the main chamber mixture.

Motivated by the above, numerical simulations were carried out using the commercial CFD code, ANSYS Fluent 15.0 [7], to simulate the flame propagation process within the pre-chamber and the penetration process of the transient turbulent hot jet issued from the pre-chamber into the main chamber. The goal was to understand why supersonic jets can extend lean flammability to achieve leaner combustion in the main chamber, which potentially can further reduce NOx emissions in gas engines. Here we chose hydrogen as the fuel due to its simple chemistry and potential as an alternative fuel. The simulations followed the exact same conditions of the experiment [6], in which six different nozzles including two straight nozzles, one converging nozzle, and three converging-diverging (C-D) nozzles were tested, and their performance was compared. For each nozzle, we examined the characteristics of the hot jet, including the spatial and temporal distributions of velocity, vorticity, Mach number, pressure, temperature, and shock structures. Additionally, the local Damköhler numbers of the gases as a function of time and location were calculated for different types of nozzles just before ignition took place in the main chamber. This comparison of local Damköhler numbers helped to reveal the fundamental mechanism of supersonic jets extending the lean flammability limit.

4.2 Experimental Method

The experimental setup is already described in detail in Chap. 2, and thus only a brief description is presented here. A small volume stainless steel pre-chamber was mounted on the top of a carbon steel main chamber. The main chamber to pre-chamber volume ratio is 100:1. A stainless steel orifice plate with various nozzle designs was placed between the two chambers to separate them. Six different nozzles were tested. Mixtures in both the chambers were initially kept at room temperature. The stoichiometric H_2/air mixture in the pre-chamber was ignited by an electric spark generated at the top of the pre-chamber. Once the spark ignited the pre-chamber mixture, the combustion products started to enter into the main chamber in the form of a hot jet which then ignited the ultra-lean main chamber. The lean limit for each nozzle was found by gradually reducing the H_2/air equivalence ratio inside the main chamber until ignition could not occur anymore. Note the H_2/air equivalence ratio of the pre-chamber mixture was fixed at $\phi = 1$ for all cases, whereas the H_2/air equivalence ratio of the main chamber mixture was varied.

4.2.1 Supersonic Nozzle Designs

A stainless steel orifice plate with various nozzle designs as shown in Fig. 4.1 separated both chambers. Jet ignition characteristics of H_2/air for six different nozzle designs, straight, convergent, and convergent-divergent (C-D), were studied. Nozzle dimensions are tabulated in Table 4.1 and schematically shown in Fig. 4.1.

4.2.2 High-Speed Schlieren and OH* Imaging

A customized trigger box synchronized with the CDI spark ignition system sent a master trigger to two high-speed cameras for simultaneous schlieren and OH*

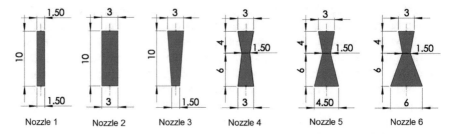

Fig. 4.1 Schematic of different types of nozzle designs. All dimensions are in millimeter [6]

Table 4.1 Nozzle type and dimensions used for the hot jet ignition experiment

Nozzle #	Type	d_{inlet} (mm)	d_{throat} (mm)	d_{exit} (mm)	A_e/A_t
1	Straight	1.5	1.5	1.5	–
2	Straight	3	3	3	–
3	Convergent	3	–	1.5	–
4	C-D	3	1.5	3	4
5	C-D	3	1.5	4.5	9
6	C-D	3	1.5	7.5	16

chemiluminescence imaging. The details of high-speed schlieren and OH* chemiluminescence systems are described in Chap. 2.

4.2.3   Hot-Wire Pyrometry and Infrared Imaging

The hot-wire pyrometry (HWP) technique provides a time-resolved temperature
field along a line during jet propagation. Planar time-dependent radiation intensity
measurements of the flame were acquired using an infrared camera (FLIR SC6100)
with an InSb detector. The view angle of the camera was aligned perpendicular to the
flame axis (50 cm from the burner center to the camera lens) such that the half view
angle of the camera is less than 10°. The radiation intensity detected by each pixel of
the camera focal plane array can be approximated by a parallel line of sight because
of the small view angle. The spatial resolution is 0.2×0.2 mm^2 for each pixel. The
band-pass filter was used to measure the radiation intensity of H_2O
(2.58 ± 0.03 μm).

4.2.4   Schlieren Image Velocimetry

In schlieren PIV (SPIV) method, a turbulent flow field containing turbulent eddies
serves as PIV particles. These self-seeded successive Schlieren images with short

time delay; Δt can be correlated to find instantaneous velocity field information. Due to path integrated nature of schlieren, an inverse Abel transformation is required to find true velocity field. A z-type Herschellian high-speed schlieren system was used for schlieren PIV. The schlieren system consisted of a 100 watt mercury arc lamp (Q series, 60,064-100MC-Q1, Newport Corporation, Model 6281) light source with a condensing lens assembly (Q Series, F/1, Fused Silica, Collimated, 200–2500 nm), two concave parabolic mirrors (6″ diameter, aperture $f/8$, effective focal length 1219.2 mm), a knife-edge, an achromatic lens ($f = 300$ mm) to collimate the light, a beam splitter (1″ cube, Thorlabs PBS251), and two identical high-speed CCD cameras (v711, Vision Research Phantom). Utilization of two high-speed cameras lies in precise controlling of the inter-frame delay, Δt. A small Δt is essential in order to resolve high exit jet velocity, U_0.

4.3 Numerical Method

4.3.1 Simulation Domain and Boundary Conditions

We aimed at computing the flame propagation process inside the pre-chamber and the penetration process of the transient hot jet until ignition occurs in the main chamber. In other words, the simulation would stop when ignition and flame propagation start in the main chamber. Figure 4.2a shows the 2D computational domain and different types of boundary conditions. Due to symmetry, only half of the domain was modeled. The computational domain was divided into three zones, namely, the pre-chamber zone, the nozzle zone, and the main chamber zone, respectively. The pre-chamber has dimensions of L (length) $\times$ D (diameter) $= 88.9$ mm $\times$ 19.05 mm. The dimensions of the nozzle connecting the two chambers are already reported in Table 4.1. The main chamber has dimensions of L (length) $\times$ D (diameter) $= 304.8$ mm $\times$ 101.6 mm. The entire domain was discretized using quadrilateral cells. Figure 4.2b shows the mesh of the entire computational domain. Figure 4.2c presents a magnified view of the mesh connecting nozzle between the pre-chamber and the main chamber. The minimum grid size is 0.05 mm. A mesh independence study was conducted by running the model on two different refined meshes – coarser and finer than the original mesh.

An axisymmetric boundary condition was used at the centerline, while everywhere else wall boundary conditions were applied. The initial wall temperature was constant at 300 K with nonslip boundary condition. Initially, the pre-chamber mixture was set at the stoichiometric condition, and the main chamber mixture was at the ultra-lean condition. At $t = 0$ the premixed H_2/air mixtures at both chambers were set to be at 300 K. The material of the pre-chamber and nozzle walls is stainless steel (SS316), and the main chamber wall material is carbon steel (C-1144) identical to the experimental conditions.

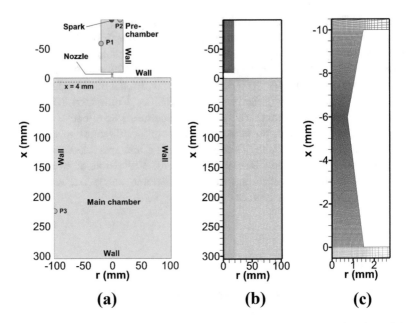

Fig. 4.2 (a) Schematic of the computational domain; (b) the computational domain and three different zones, namely, pre-chamber, nozzle, and main chamber zones; and (c) the magnified view of the mesh of the connecting nozzle

At the beginning of the simulation, a spark with an energy of 100 mJ was supplied at the location (2, 0 mm) within the pre-chamber to initiate ignition. At first, a laminar semispherical flame was observed to propagate outwardly. As the pressure in the pre-chamber was building up, part of the unburned gases in the pre-chamber was pushed out into the main chamber. Eventually, the hot combusted products from the pre-chamber combustion were pushed out into the main chamber in the form of a turbulent hot jet. The simulations stopped at the instance just before ignition started in the main chamber. This time, when the simulation would stop, was measured experimentally and fed into the numerical model. The simulations did not include the ignition and combustion processes in the main chamber, and the focus was on the characteristics of the turbulent hot jet.

4.3.2   Governing Equations

The governing equations that describe transient chemically reacting compressible flows are the unsteady Reynolds averaged Navier-Stokes (U-RANS) equations coupled with mass, energy, and species conservation equations. There are $(N - 1)$

species conservation equations, where N is the total number of species. These partial differential equations can be written as

$$\frac{\partial \rho}{\partial t} + \frac{\partial}{\partial x_j}(\rho u_j) = 0 \tag{4.1}$$

$$\frac{\partial}{\partial t}(\rho u_i) + \frac{\partial}{\partial x_j}(\rho u_i u_j + \delta_{ij}P - \sigma_{ij}) - \rho g_j = 0 \tag{4.2}$$

$$\frac{\partial}{\partial t}(\rho \mathbb{E}) + \frac{\partial}{\partial x_j}(\rho \mathbb{H} u_j + q_j - \sigma_{ij}u_i) = 0 \tag{4.3}$$

$$\frac{\partial}{\partial t}(\rho Y_m) + \frac{\partial}{\partial x_j}(\rho Y_m u_j + \rho Y_m V_j) = \dot{w}_m \tag{4.4}$$

where t is time, ρ is the density, P is the pressure, u is the velocity, $\mathbb{E}$ is the total energy, $\mathbb{H}$ is the total enthalpy, σ_{ij} is the stress tensor, q_j is the heat flux vector, δ_{ij} is the Kronecker delta, and Y_m, V_j and $\dot{w}_m$ are the mass fraction, diffusion velocity, and the production rate of the mth species, respectively. The time-averaged conservation equations are obtained by decomposing each flow variable, $\rho, u_i, P, \sigma_{ij}, \mathbb{E}, \mathbb{H}, q_i, Y_m, \dot{w}_m$, into a mean and fluctuating part and then plugging into Eqs. (4.1), (4.2), (4.3) and (4.4). A mass-weighted average, known as the Favre average, is used to decompose the dependent variable f into a mean $\tilde{f}$ and a fluctuating f'' part.

$$f = \tilde{f} + f'' \tag{4.5}$$

$$\tilde{f} = \frac{1}{\rho} lim_{\Delta t \to \infty} \frac{1}{\Delta t} \int_{t_0}^{t_0 + \Delta t} \rho f dt \tag{4.6}$$

Substituting the decomposed variables into Eqs. (4.1), (4.2), (4.3) and (4.4) and using Favre averaging as shown in Eq. (4.6) yield the desired time-averaged equations:

$$\frac{\partial \bar{\rho}}{\partial t} + \frac{\partial}{\partial t}(\bar{\rho} \tilde{u}_j) = 0 \tag{4.7}$$

$$\frac{\partial}{\partial t}(\bar{\rho} \tilde{u}_i) + \frac{\partial}{\partial t}(\bar{\rho} \tilde{u}_i \tilde{u}_j + \delta_{ij}\bar{P}) = \frac{\partial}{\partial x_j}(\bar{\sigma}_{ij} - \overline{\rho u_i'' u_j''}) \tag{4.8}$$

$$\frac{\partial}{\partial t}(\bar{\rho} \tilde{\mathbb{E}}) + \frac{\partial}{\partial x_j}(\bar{\rho} \tilde{\mathbb{H}} \tilde{u}_j) = \frac{\partial}{\partial x_j}(\bar{\sigma}_{ij}\tilde{u}_i + \overline{\sigma_{ij}u_i''} - \bar{q}_j - \overline{\rho \mathbb{H}'' u_j''}) \tag{4.9}$$

$$\frac{\partial}{\partial t}(\bar{\rho} \tilde{Y}_m) + \frac{\partial}{\partial x_j}(\bar{\rho} \tilde{Y}_m \tilde{u}_j) = \dot{w}_m - \frac{\partial}{\partial x_j}(\overline{\rho Y_m V_j} + \overline{\rho Y_m'' u_j''}) \tag{4.10}$$

All the terms on the right-hand side of Eqs. (4.7), (4.8), (4.9) and (4.10) require modeling assumptions. The time-averaged molecular stress tensor, $\bar{\sigma}_{ij}$, as shown in Eqs. (4.8) and (4.9) can be modeled by ignoring the effects of turbulent fluctuations on the molecular viscosity, μ, as

$$\overline{\sigma}_{ij} = \mu \left(\frac{\partial \tilde{u}_i}{\partial x_j} + \frac{\partial \tilde{u}_j}{\partial x_i} \right) - \frac{2}{3} \delta_{ij} \mu \frac{\partial \tilde{u}_k}{\partial x_k} \tag{4.11}$$

The second term in right-hand side of the decomposed energy equation, $\overline{\sigma_{ij} u_i''}$, can be correlated with the mean turbulent kinetic energy, $\tilde{k} = \widehat{u_i'' u_i''}/2$, using the following approximation:

$$\frac{\partial}{\partial x_j} \left(\overline{\sigma_{ij} u_i''} \right) = \frac{\partial}{\partial x_j} \left(\frac{\mu \partial \tilde{k}}{\partial x_j} \right) \tag{4.12}$$

The time-averaged heat flux vector, $\overline{q}_j$, usually, contains contributions from heat conduction and an energy flux term due to interspecies diffusion. $\overline{q}_j$ can be written as

$$\overline{q}_j = -\lambda \frac{\partial \tilde{T}}{\partial x_j} + \sum_{m=1}^{N} \overline{\rho Y_m V_j h_m(T)} \tag{4.13}$$

The total energy, $\mathbb{E}$, is the sum of the internal energy, $e = c_v T$, and the kinetic energy, $k = u_i u_i/2$. The total enthalpy is the sum of total energy, $\mathbb{E}$, and P/ρ. Pressure, P, is given by the ideal gas equation, $P = \rho RT$, where $R = R_u \sum\limits_{m=1}^{N} Y_m / W_m$ is the gas constant, W_m is the molecular weight of the *mth* species, and R_u is the universal gas constant. Thus, the total enthalpy, $\mathbb{H}$, can be expressed as $\mathbb{H} = h + u_i u_i/2 + P/\rho$. The diffusion velocity of the *mth* species is usually evaluated from Fick's law of diffusion and can be expressed in the following form:

$$V_j = -\frac{D_m}{Y_m} \frac{\partial Y_m}{\partial x_j} \tag{4.14}$$

Here, D_m is the mass diffusivity of the *mth* species relative to the mixture.

These governing equations, unsteady Reynolds averaged Navier-Stokes (U-RANS) equations, were solved in the 2D axisymmetric domain as shown in Fig. 4.2 using the commercial code ANSYS Fluent R15.0. Each simulation was run in parallel on 4 nodes; each node contained 16 processors; a total of 64 processors were used per simulation. The computational time for each simulation is about 6–7 days to obtain a simulation time of approximately 50 ms. Details of the turbulence modeling, turbulence-chemistry interaction, and numerical solver setting are discussed in the subsequent sections.

4.3.3 Turbulence Modeling

The Reynolds stress models (RSM) [8], a higher-level turbulence model, was used to calculate individual Reynolds stresses, $\overline{\rho u_i'' u_j''}$. RSM has the superior competency to predict complex flows accurately as it accounts for the effects of swirl, rotation, streamline curvature, and rapidly changing strain rates in a more rigorous manner than other turbulence models like $k - \epsilon$ [9, 10] or $k - \omega$ [11]. However, the accuracy of RSM models is limited by the closure assumptions employed to model various terms such as the turbulent diffusive transport and pressure-strain terms in the Reynolds stress transport equation. A gradient-diffusion model of Daly and Harlow [12] was used to model turbulent diffusive transport. A pressure-strain model proposed by Speziale, Sarkar, and Gatski [13] was chosen due to its superior performance for axisymmetric expansion and contraction.

4.3.4 Chemistry Modeling

A detailed chemistry model including 9 species and 21 reactions [14] was used for H_2/air in the present simulations. The turbulence-chemistry interaction was modeled using the eddy dissipation concept (EDC) [15] model. The EDC model assumes that reaction occurs in small turbulent structures, called the fine scales. This model has the capability to accurately include detailed chemical reaction mechanisms.

A burning velocity model is necessary to calculate the burning velocities of H_2/air at different locations and time steps in the simulation. The laminar burning velocity of H_2/air mixture depends on the equivalence ratio, pressure, and temperature. Based on the burning model proposed by Iijima and Takeno [16], the laminar burning velocity, $s_L(T, P)$, of H_2/air can be expressed as

$$\frac{s_L(T,P)}{s_L^0(T_0,\ P_0)} = \left(\frac{T}{T_0}\right)^{\alpha_1}\left[1 + \alpha_2\ln\left(\frac{P}{P_0}\right)\right] \quad (4.15)$$

where $s_L^0(T, P)$ denotes the laminar burning velocity at the reference condition T_0 and P_0, $\alpha_1 = 1.54 + 0.026(\phi - 1)$, and $\alpha_2 = 0.43 + 0.003(\phi - 1)$. Both α_1 and α_2 have a weak dependence on the equivalence ratio, ϕ.

4.3.5 Numerical Details

The compressible Navier-Stokes equations were solved using a pressure-based solver in which the pressure and velocity were coupled using the SIMPLE [17] algorithm. At the beginning of the simulation for a few milliseconds, a first-order upwind discretization scheme was used for the convective terms and turbulent quantities to obtain a stable, first-order accurate solution. Once a stable solution was reached, we switched the discretization scheme to third-order MUSCL [18] for an accurate solution. However, this higher-order discretization scheme increased computation time significantly. The least squares cell-based gradient calculation scheme, which is known for accuracy and yet computationally less expensive, was chosen over the node-based gradient for the spatial discretization. A second-order discretization scheme was used for pressure. The solution-adaptive mesh refinement feature was used to resolve shock structures. A dynamic adaption of the pressure gradient was implemented to refine the mesh near the shock or to coarsen it wherever needed. A fixed time step of $t = 10^{-5}$ second was used to resolve the chemical timescale. The second-order implicit scheme was used for time integration of each conservation equation.

4.4 Results and Discussions

4.4.1 Experimental Results

Results and discussions are divided into following subsections. At first, lean limit and ignition delay for different types of nozzles are discussed, followed by jet ignition mechanism using simultaneous schlieren and OH* chemiluminescence imaging. Next supersonic jet characteristics, shock structures at the jet exit, and qualitative temperature field measurements are presented using simultaneous schlieren and IR imaging. Radiation intensity field from infrared diagnostics and temperature profile near jet exit was measured using hot-wire pyrometry (HWP).

4.4.1.1 Lean Flammability Limit and Ignition Delay

One of the main goals was to understand the effect of nozzle geometry on the lean ignition limit in the main chamber. Six different nozzles were tested; their dimensions are summarized in Table 4.1. The lean limit for each nozzle was found by gradually reducing the fuel equivalence ratio of the main chamber until ignition cannot occur. Note the fuel/air equivalence ratio of the pre-chamber mixture was fixed to $\phi = 1$ for all cases, whereas the fuel/air equivalence ratio of the main chamber mixture was varied. Figure 4.3 show the lean limit (ϕ_{limit}) of the main chamber mixture for six nozzles. As can be seen, ϕ_{limit} extends for supersonic nozzles compared to its straight counterpart. Out of the four supersonic nozzles we tested, nozzle 4 and nozzle 5 showed lowest lean limit, $\phi_{limit} = 0.22$ and 0.23, respectively.

Supersonic Nozzles also show a lower ignition delay time, $\tau_{ignition}$, compared to its straight counterpart. The ignition delay, $\tau_{ignition}$, as a function of the equivalence ratio, ϕ, is plotted in Fig. 4.4. As mentioned earlier, ignition delay is the time required for the initiation of main chamber ignition from the diaphragm rupture. Ignition delay is smaller for supersonic nozzles 4 and 5 compared to straight nozzles 1 and 2.

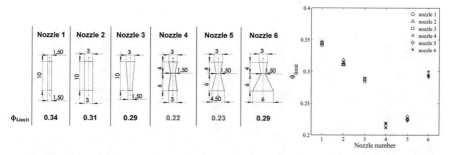

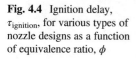

Fig. 4.3 Lean limit ϕ_{limit} for various types of nozzles. Supersonic nozzles, nozzle 4 and 5, show lowest equivalence ratio, ϕ_{limit}, of 0.22 and 0.23, respectively, compared to straight nozzles

Fig. 4.4 Ignition delay, τ_{ignition}, for various types of nozzle designs as a function of equivalence ratio, ϕ

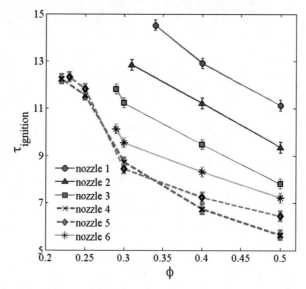

4.4.1.2 Visualization of Ignition Processes in Main Chamber

The high-speed schlieren technique enables visualization of the jet penetration process, as well as the ignition and subsequent turbulent flame propagation processes in the main chamber. High-speed OH* chemiluminescence helps to identify the flame front location and the ignition mechanism (whether the hot jet coming out from the pre-chamber is a jet of hot combustion products or a jet of flames). For all ultra-lean cases, ignition occurred via jet ignition mechanism (a jet consists of only hot combustion products, which resulted in ignition of the ultra-lean mixture in the main chamber). In our previous studies with the main chamber equivalence ratio at or near stoichiometry ($\phi = 0.75 - 1$), we observed ignition could occur via flame jet ignition mechanism as well. Unlike a hot jet of combustion products, a flame jet produces a jet full of wrinkled turbulent flames and active radicals. Initial pressure,

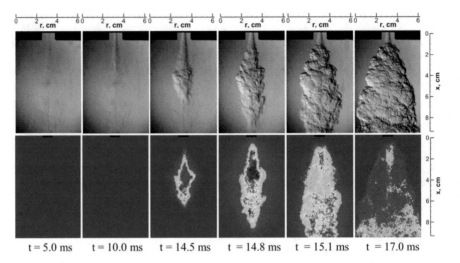

t = 5.0 ms   t = 10.0 ms   t = 14.5 ms   t = 14.8 ms   t = 15.1 ms   t = 17.0 ms

Fig. 4.5 The time sequence of simultaneous schlieren (top) and OH* chemiluminescence (bottom) images showing jet ignition process for *nozzle 1*. Experimental conditions: $V_{pre-chamber} = 100$ cc, $P_{initial} = 0.1$ MPa, $T_{initial} = 300$ K, $\phi_{pre-chamber} = 1.0$, $\phi_{main\ chamber} = 0.34$. The ignition delay, $\tau_{ignition}$, is 14.52 ms

temperature, equivalence ratio along with geometric factors like pre-chamber volume, orifice diameter, and spark position all affect ignition behavior and determine which ignition mechanism will dominate. At lower equivalence ratios and lower pressures, jet ignition mechanism is predominant. Higher initial temperatures or pressures lead to flame ignition mechanism. Figures 4.5, 4.6, 4.7, 4.8, 4.9, and 4.10 show the time sequence of simultaneous schlieren (top) and OH* chemiluminescence (bottom) of the ignition process for different nozzles at the lean-limit condition.

Figures 4.5 and 4.6 represent jet ignition by straight nozzles of diameters 1.5 and 3 mm, respectively. Initially, shock structures are visible for a little while, but these structures diminish quickly, and ignition starts afterward in the absence of any shock structures. As for convergent and C-D nozzles, jet contains straight/diamond shock structures at the exit as seen from Figs. 4.7, 4.8, 4.9 and 4.10, and ignition starts at the presence of shock structures. With increase in area ratio, A_e/A_t jet width at nozzle exit increases as well. Figures 4.8 and 4.9 show the ignition processes for C-D nozzles with area ratio of 4 and 9, respectively. ϕ_{limit} for these two nozzles are 0.22 and 0.23, respectively, and are the smallest compared to other nozzles.

In following numerical simulation section, we will show that the shock structures in the supersonic jets increase static temperature behind the shocks, which in turn increase the ignition probability of the main chamber mixture and thus reduce the lean limit. This is also depicted as in increase in radiation intensity downstream of nozzle exit by IR imaging discussed in the following section.

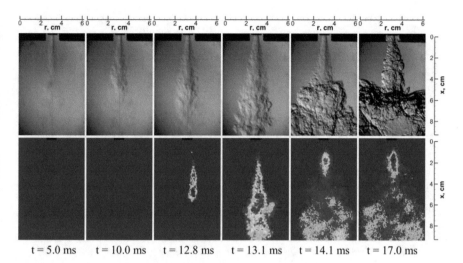

Fig. 4.6 The time sequence of simultaneous schlieren (top) and OH* chemiluminescence (bottom) images showing jet ignition process for *nozzle 2*. Experimental conditions: $V_{pre-chamber} = 100$ cc, $P_{initial} = 0.1$ MPa, $T_{initial} = 300$ K, $\phi_{pre-chamber} = 1.0$, $\phi_{main\ chamber} = 0.31$. The ignition delay, $\tau_{ignition}$, is 12.84 ms

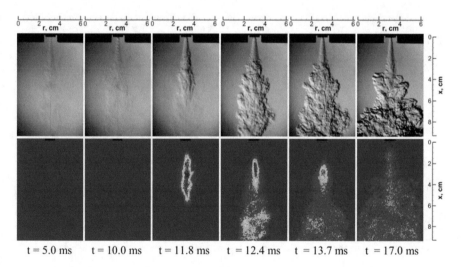

Fig. 4.7 The time sequence of simultaneous schlieren (top) and OH* chemiluminescence (bottom) images showing jet ignition process for *nozzle 3*. Experimental conditions: $V_{pre-chamber} = 100$ cc, $P_{initial} = 0.1$ MPa, $T_{initial} = 300$ K, $\phi_{pre-chamber} = 1.0$, $\phi_{main\ chamber} = 0.29$. The ignition delay, $\tau_{ignition}$, is 11.82 ms

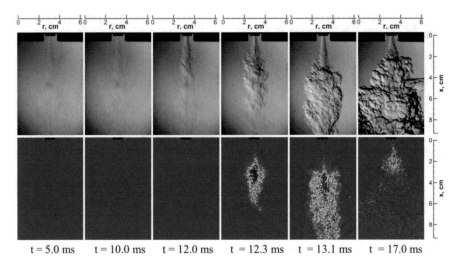

Fig. 4.8 The time sequence of simultaneous schlieren (top) and OH* chemiluminescence (bottom) images showing jet ignition process for *nozzle 4*. Experimental conditions: $V_{pre-chamber} = 100$ cc, $P_{initial} = 0.1$ MPa, $T_{initial} = 300$ K, $\phi_{pre-chamber} = 1.0$, $\phi_{main\ chamber} = 0.22$. The ignition delay, $\tau_{ignition}$, is 12.24 ms

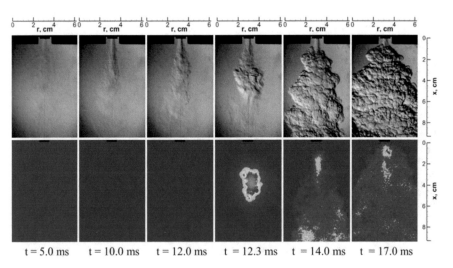

Fig. 4.9 The time sequence of simultaneous schlieren (top) and OH* chemiluminescence (bottom) images showing jet ignition process for *nozzle 5*. Experimental conditions: $V_{pre-chamber} = 100$ cc, $P_{initial} = 0.1$ MPa, $T_{initial} = 300$ K, $\phi_{pre-chamber} = 1.0$, $\phi_{main\ chamber} = 0.23$. The ignition delay, $\tau_{ignition}$, is 12.32 ms

4.4.1.3 Infrared Measurements

Figure 4.11 shows simultaneous planar time-dependent radiation intensity measurements and high-speed schlieren imaging to capture shock structure of turbulent hot

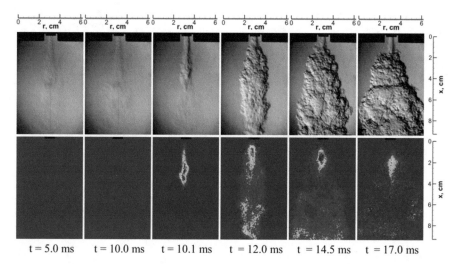

Fig. 4.10 The time sequence of simultaneous schlieren (top) and OH* chemiluminescence (bottom) images showing jet ignition process for *nozzle 6*. Experimental conditions: $V_{pre-chamber} = 100$ cc, $P_{initial} = 0.1$ MPa, $T_{initial} = 300$ K, $\phi_{pre-chamber} = 1.0$, $\phi_{main\ chamber} = 0.29$. The ignition delay, $\tau_{ignition}$, is 10.12 ms

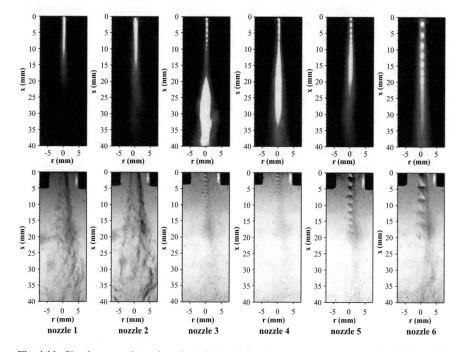

Fig. 4.11 Simultaneous planar time-dependent radiation intensity measurements of the hot jet with H_2O (2.58 ± 0.03 μm) band-pass filter and high-speed schlieren imaging reveal the shock structure of supersonic jets. The high-temperature zone downstream of the shock structure is marked by a yellow box

Fig. 4.12 Radial
temperature profile
measured by HWP
technique at $x = 4$ mm
downstream of the
nozzle exit

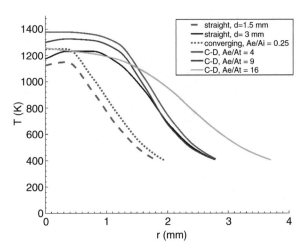

jets. These images represent the time instance just before the ignition occurs in the
main chamber. The main chamber was filled with only air (nonreactive) to characterize the hot jet. A qualitative understanding of the temperature field can be obtained
from the infrared images. Perhaps the most significant feature of these infrared
images is the presence of shock structures for converging and C-D nozzles at the
instance of main chamber ignition. Unlike convergent or C-D nozzles, straight
nozzles do not show any shock structures just before ignition. Besides shock
structures, a high-temperature zone at the downstream of the shock structure was
observed for convergent and C-D nozzles. This is due to the fact that static temperature rises after each shock and creates a high-temperature zone. The location and
width of the zone vary for different nozzles. But the much interesting fact lies in the
ignition pattern of these nozzles. For all these nozzles, ignition starts from these
high-temperature zones.

The jet temperature at a location that is 4 mm downstream of the nozzle exit was
measured using hot-wire pyrometry (HWP) technique. The results are shown in
Fig. 4.12. Nozzles 4 and 5 exhibit higher temperatures at the centerline than the other
nozzles. Figure 4.13 reveals the radiation intensity along the jet centerline in the
streamwise direction.

For straight nozzles (nozzle 1 and 2), the radiation intensity drops in a monotonic
fashion, indicating the temperature of the jet keep decreasing as a result of mixing
between the hot jet and the cold ambient mixture. However, for nozzles 3, 4, and
5, the measured radiation intensity first fluctuates near the nozzle exit due to the
presence of shocks, for which the static temperature increases downstream of the
shock. It then increases rapidly at a location further downstream, indicating the
establishment of a higher-temperature zone at that location. Resulted ignition of the
main chamber lean mixture was observed to take place in this high-temperature zone
for nozzles 3, 4, and 5. In other words, this high-temperature zone downstream of the
nozzle exit is responsible for reducing the lean limit of the main chamber mixture by
using a supersonic nozzle.

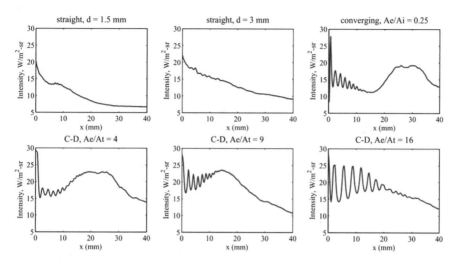

Fig. 4.13 Measured instantaneous radiation intensity at the jet centerline for different types of nozzle design

4.4.2 Numerical Results

Numerical results from six different nozzle geometries are presented here. Quantitative comparison of different geometry nozzles was performed to explain the superior performance of supersonic jets over subsonic jets. At first, the flame propagation process within the pre-chamber is discussed in detail. Then the spatial variation of velocity, vorticity, shock, Mach number, species concentration, and temperature are presented and discussed in depth. Lastly, the local Damköhler numbers are calculated based on the local flow properties just before ignition in the main chamber, which helps to understand the physics of ignition process by supersonic hot jets.

4.4.2.1 Validation

In the experiment [6], the pressure histories of both chambers were recorded using high-speed pressure transducers. Additionally, the temperature of the jet at several locations downstream of the nozzle exit was measured using the hot-wire pyrometry (HWP) technique. Here we have compared the computed and measured pressure histories of the two chambers as well as the temperature profiles of the jet with the intent to provide a validation of the computational method.

As depicted in Fig. 4.2a, in the numerical simulations, the pressure history of the pre-chamber was monitored at two locations P_1 and P_2, while the main chamber pressure history was monitored at location P_3. These locations were chosen because they were the same locations where the pressure transducers were installed in the

Fig. 4.14 Comparison of
the measured and computed
pre-chamber pressure
histories

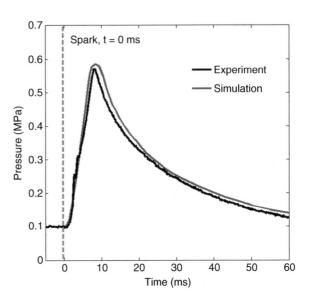

experiment. Figure 4.14 compares the measured and computed pressure histories in
the pre-chamber, which agree well. Both show that after a short ignition delay, the
pre-chamber combustion initiated and the pressure started rising. The peak pressure,
which is almost six times the initial pressure, occurred at about 9 ms after ignition,
indicating combustion in the pre-chamber had been completed by that time. Afterward, pressure started to drop because the combustion products had been pushed into
the main chamber. The computed pressure around the peak is slightly higher than the
experimentally measured values, which may be because heat loss (from the
pre-chamber to the surroundings) was neglected in the simulations.

Figure 4.15 compares the measured and computed jet temperature at a location
4 mm downstream of the nozzle exit for all six nozzles. The temperature is highest at
the centerline, then decays in the mixing layer, and finally, reaches the ambient
temperature. Among the six nozzles, nozzle 4 results in the highest centerline
temperature followed by nozzles 5 and 6. The agreement between measurements
and simulations is good. However, the computed temperatures are higher than the
measured values in the regions around the jet centerline and the mixing layers.

4.4.2.2 Flame Propagation Process in the Pre-chamber

Flame propagation in a small volume such as a pre-chamber is more complicated
than in a large volume. This is because flame propagation is constrained by walls/
boundaries, which may induce instabilities, acoustic waves, vortex flow, and flame
acceleration and deceleration [19–21]. For example, a flame propagating in a small
tube can experience shape changes, from spherical, curved, to the tulip and cellular
fronts [22, 23]. The aspect ratio (length to diameter) of the tube is a crucial

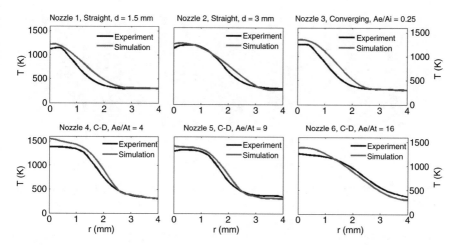

Fig. 4.15 Comparison of computed and measured temperature profiles of the hot jet at $x = 4$ mm downstream of the nozzle exit

parameter. A tulip flame was observed in a closed tube when the aspect ratio is greater than 2 [21, 24]. Additionally, unlike most studies in the literature that used a tube with either an open end or a closed end, the pre-chamber configuration has a small opening at the end wall, connecting the two chambers. As such, the flame propagation process is also influenced by this opening, which allows gases flowing into the main chamber and thus reduces pre-chamber pressure. In the following, we will discuss the flame propagation process in the pre-chamber.

Figure 4.16 shows the transient flame propagation process in the pre-chamber filled with stoichiometric premixed H_2/air. Figure 4.17 presents the calculated propagation speed (displacement speed) of the leading flame tip and the flame surface area inside the pre-chamber as a function of time.

After an ignition spark had taken place at the top center of the pre-chamber at $t = 0$, the flame developed a hemispherical shape and was expanding outwardly at a velocity close to the laminar flame speed of stoichiometric H_2/air. At this stage, the flame propagated freely at almost a constant speed. Soon after (0.1 ms $< t < 0.55$ ms), this hemispherical flame started accelerating with an exponential increase in the flame surface area due to the confinement of the pre-chamber sidewalls. This is consistent with the observations of Xiao [25–27], Ponizy [28], and Markstein [29]. For example, Xiao investigated the dynamics of premixed H_2/air flame in a closed combustion vessel. They observed the flame propagated nearly at a constant speed when it was far away from the wall/boundaries. The flame went through a fast acceleration as the flame surface area was increased. As it approached the wall/boundaries, the flame propagation speed dropped. At a later stage, the flame propagation speed increased again because of the formation of the tulip shape. Such trend was also observed for flame propagation in the pre-chamber in the present study.

As shown in Figs. 4.16 and 4.17, when the expanding, hemispherical flame front finally touched the pre-chamber sidewalls, the flame stopped accelerating. The

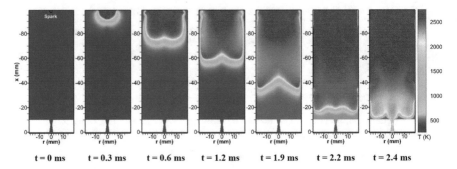

Fig. 4.16 The computed flame propagation process in a stoichiometric H$_2$/air mixture in the pre-chamber

Fig. 4.17 The propagation speed of the leading flame tip and the flame surface area as a function of time

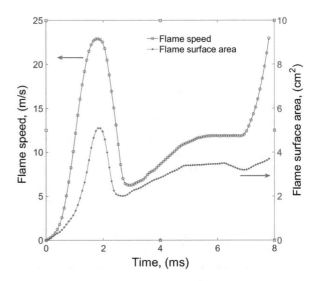

deceleration was caused by the significantly reduced flame surface area as well as heat losses from the flame to the walls. After $t = 0.55$ ms, the flame front started changing into a tulip shape. Several theories have been proposed to explain this transition of flame structure from semispherical to growing concave, e.g., the effect of Darrius-Landau and Taylor instability [28], interaction of the flame front with its self-generated pressure waves [50], viscous flow interaction with the flame front [30], and the vortical fluid flow interaction with the flame front [19, 20, 24]. The formation of the tulip shape led to a second increase of flame surface area. Thus, the flame accelerated again after the inversion in the duration of 1 ms $< t <$ 1.7 ms. Between 1.7 ms $< t <$ 2.2 ms the flame surface area remained almost unchanged. Hence, the flame propagation speed remained constant as well. The last stage of the pre-chamber flame propagation occurred in the range of 2.2 ms $< t <$ 2.4 ms when the tulip-shaped flame was approaching the connecting nozzle inlet. The concave flame front at the centerline started accelerating forming a convex shape before entering the connecting nozzle.

4.4.2.3 Characteristics of the Hot Jet

In our previous experiment [5], we found that when the nozzle diameter was small enough, the pre-chamber flame extinguished while passing through the nozzle due to the high stretch rate and heat losses through the walls. As a result, the jet entering the main chamber contained hot combustion products only. Very little intermediate species and radicals were present in the jet, as demonstrated by high-speed OH chemiluminescence imaging. Additionally, main chamber ignition did not occur as soon as the hot jet penetrated into the main chamber. The hot jet first accelerated and then decelerated. Main chamber ignition, however, always took place during the jet deceleration process. Furthermore, ignition usually took place from the side surface of the hot jet at a location downstream of the nozzle exit. In the most recent experiment [6], we found that using C-D nozzles the mixture in the main chamber could be leaner than using straight nozzles. Because the probability of ignition is fundamentally determined by the Damköhler number (the competition between turbulent mixing and ignition chemistry), we must, first, understand the characteristics of the hot jet, particularly the mixing process between the hot jet and the cold ambient. Motivated by this, we compared the characteristics of the hot jet using six different nozzles, including profiles of velocity, Mach number, vorticity, species concentration, pressure and temperature distributions, and shock structures for the C-D nozzles.

Mach Number

Figure 4.18 shows the spatial distribution of Mach number profiles for all six nozzles just before ignition occurred in the main chamber at the lean flammability limit of the corresponding nozzles as reported in Table 4.1. Initially, for all the test conditions, both the main chamber and pre-chamber were at 1 atm and 300 K. The equivalence ratio in the pre-chamber was at stoichiometric for all the tests, while the main chamber equivalence ratio was at the lean flammability limit of the corresponding nozzle. We measured the time, t_S, between spark initiation in the pre-chamber and ignition in the main chamber in our previous experiment [6] for the identical test conditions. This time, t_S, was used for the simulations, after which the simulations stopped. In other words, the simulations did not include the ignition and flame propagation process in the main chamber because our purpose was to examine the jet characteristics immediately before main chamber ignition. For the straight nozzles (nozzle 1 and nozzle 2), the jet remained subsonic when the ignition started in the main chamber. However, for the other nozzles, the jet was supersonic and exhibited complex shock structures. The converging nozzle (nozzle 3) showed under-expanded flow pattern, while the three C-D nozzles showed overexpanded flow patterns. With an increase in the area ratio of the C-D nozzles, from 4 to 16, the exit Mach number became even more supersonic. Table 4.2 lists the exit Mach number at the centerline for all six nozzles before ignition in the main chamber.

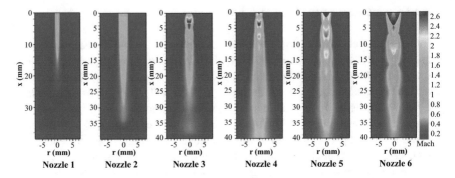

Fig. 4.18 Spatial distribution of Mach number profiles at the instance immediately before ignition initiated in the main chamber at the lean flammability limit of corresponding nozzles as shown in Table 4.1

Table 4.2 Exit Mach number at the nozzle centerline before main chamber ignition

Nozzle #	Exit Mach
Nozzle 1, straight, $d = 1.5$ mm	0.91
Nozzle 2, straight, $d = 3$ mm	0.80
Nozzle 3, convergent, $A_i/A_e = 4$	1.92
Nozzle 4, C-D, area ratio $= 4$	2.63
Nozzle 5, C-D, area ratio $= 9$	2.75
Nozzle 6, C-D, area ratio $= 16$	2.91

Shock Structures

The spatial distribution of pressure (top) and shock (bottom) profiles for all six nozzles is shown in Fig. 4.19 just before ignition took place in the main chamber. Pressure distribution is nearly uniform for the two straight nozzles. It, however, varies spatially from 0.05 MPa to 0.4 MPa for the supersonic nozzles. A closer look at the jet exit shows that the converging nozzle (nozzle 3) created an under-expanded flow, since the exit pressure, p_e, was greater than the ambient pressure, p_a. For all the C-D nozzles, the exit pressure was less than ambient pressure, indicating overexpanded flows. Additionally, based on the spatial distribution of pressure, we can clearly see the shock diamonds for the supersonic nozzles. It is well known that shock diamonds are formed due to the abrupt change in local density and pressure caused by complicated interactions between oblique shock waves and expansion fans. The size of the diamond increases as the flow Mach number increases. Nozzle 6, which has the largest area ratio of 16, exhibited the biggest shock diamonds. The static pressure and temperature changed across these diamonds and, in turn, changed the mixing between the hot jet and the cold ambient unburned fuel/air mixture. The characteristics of the shock diamonds will be discussed in detail in the following paragraphs.

Figure 4.19 (bottom) shows the computed shock structures from different nozzles. Note that numerical dissipation and oscillation in the CFD code might cause

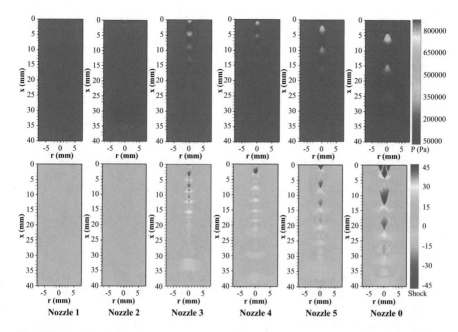

Fig. 4.19 Spatial distribution of pressure (top) and shock (bottom) profiles just before ignition in the main chamber at the lean flammability limit of corresponding nozzles as shown in Table 4.1

some shock waves to be undetected. Also, numerical oscillations might produce structures as if they were shocks and thus might lead to false shock detection. Due to these considerations, an advanced mesh adaptation technique was used to refine the mesh based on static pressure gradient information. In the present study, shocks were detected based on the method proposed by Lovely and Haimes [31]. The strength of the shock $\mathbb{S}$ can be calculated as

$$\mathbb{S} = \frac{\overrightarrow{Ma} \cdot \nabla P}{|\nabla P|} = \frac{\vec{u} \cdot \nabla P}{a|\nabla P|} \tag{4.16}$$

where $\overrightarrow{Ma}$ is the Mach number vector in the direction of the local flow velocity $\vec{u}$, a is the local speed of sound, and P is the pressure field.

As expected, the shock structures are only visible for the supersonic nozzles. After every shock diamond or shock cell, there existed a normal shock in the jet flow called Mach disk. The location of the Mach disk is important because it affects the static temperature distribution of the jet. Since centerline static temperature rises after every Mach disk, the position of the Mach disk and the size of the shock diamond or shock cell are crucial. Table 4.3 shows the position of the first visible Mach disk and the size of the first shock cell/diamond. The position of the Mach disk matches well with the theoretical approximation [32]

Table 4.3 Position of the first visible Mach disk and the size of the first shock cell/diamond

Nozzle #	First disk position, mm theoretical (Eq. (4.17) [32])	First disk position, mm computational	First cell size, mm
Nozzle 3, convergent, A_i/ $A_e = 4$	3.25	3.2	4
Nozzle 4, C-D, area ratio = 4	0.82	0.8	3.9
Nozzle 5, C-D, area ratio = 9	2.24	2.2	7.1
Nozzle 6, C-D, area ratio = 16	4.64	4.5	11.5

$$x = \frac{2}{3} d_{\text{exit}} \sqrt{\frac{P_0}{P_b}} \qquad (4.17)$$

as shown in Table 4.3. Here x is the distance between the nozzle exit and the Mach disk, d_{exit} is the nozzle diameter, P_0 is the pre-chamber pressure, and P_b is the back pressure.

The position of the first Mach disk shifted downstream with increased area ratio. The shock diamond/cell size increased with an increase in area ratio as well. However, for a particular area ratio, the shock diamond/cell size decreased slightly in the downstream. Since we have discovered from our previous experiments [6] that ignition occurred from a high-temperature zone after the final strong shock, Table 4.2 implies that for a higher area ratio nozzle, main chamber ignition would start from a further downstream location of the jet.

Figure 4.20 shows the detailed shock structure of two flow patterns from two different nozzles, nozzle 3 and nozzle 5, respectively, just before ignition occurred in the main chamber. Nozzle 3 produced an under-expanded flow, while nozzle 5 produced an overexpanded flow. Like nozzle 5, nozzles 4 and 6 also produced overexpanded flows. The variation in the shock structure depends on the pre-chamber pressure and the back pressure, p_b, in the main chamber. With sufficient increase in pre-chamber pressure, a shock is generated just after the flow passed the throat of the nozzle and moved downstream. If $P_e < P_a$, the shock starts to compress inward, in the form of an oblique shock. As the pre-chamber pressure increases during combustion, the hot turbulent jet is unable to adjust to the back pressure inside the nozzle but rather adjusts inside the main chamber in the form of compression waves or expansion waves as shown in Fig. 4.20. As the hot jet enters from the converging nozzle to lower pressure surroundings (under expansion $P_e > P_a$), it follows Prandtl-Meyer expansion at the exit of the nozzle. These expansion waves are then reflected from the constant pressure jet boundary as compression waves as shown in Fig. 4.20. The compression waves intersect each other when the exit pressure is higher than ambient pressure. This lead to shock waves being formed

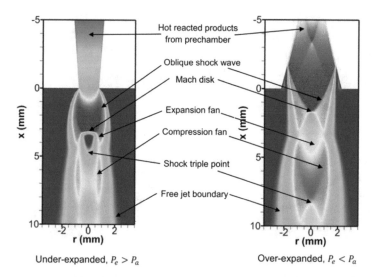

Fig. 4.20 Detailed shock structures of under-expanded (nozzle 3) and overexpanded (nozzle 4) nozzle flows just before ignition in the main chamber at the lean flammability limit

in the jet because of coalescence of the compression waves. It creates a Mach disk and the triple point where Mach disk interests with the oblique shock waves. These shock patterns gradually started to diminish further downstream of the nozzle as viscous dissipation effects along the free stream jet boundary cease the generation of further shocks along the jet boundary layer [33–35].

Velocity and Vorticity

Figure 4.21 plots the velocity profiles of the jet at four downstream locations ($x = 10$, 20, 30, and 40 mm) for six different nozzles at their corresponding lean flammability limit. These velocity profiles were just prior to the main chamber ignition and corresponded to the Mach profiles shown in Fig. 4.18. Comparing the two straight nozzles, the jet produced by nozzle 1 has higher centerline velocity than that by nozzle 2, due to the smaller diameter of nozzle 1. Nozzles 1 and 2 both showed a top-hat velocity profile at the near and intermediate region of the jet. For all supersonic nozzles, a drop in the centerline velocity could be observed at $x = 10$ mm, due to the presence of the shocks. The drop in velocity at the centerline, however, vanished at the downstream locations $x > 20$ mm.

For the converging and C-D nozzles, the centerline velocity of the jets at $x = 10$ mm was near twice the value of the jet produced by the straight nozzles. Nozzles 4 and 6 had the highest velocities at $x = 10$ mm. As the jet started spreading downstream, the jet velocities went down. At $x = 10$ mm, the velocity of nozzle

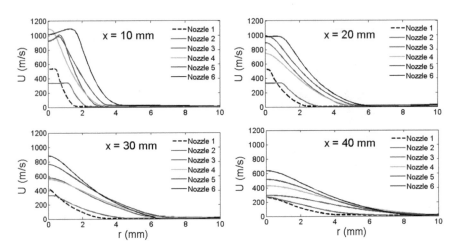

Fig. 4.21 Velocity distributions at different axial locations, $x = 10$, 20, 30, and 40 mm, just before ignition occurred in the main chamber at lean flammability limit of corresponding nozzles as shown in Table 4.1

6 was highest among the supersonic nozzles. The jet spread nearly doubled from $x = 10$ mm to $x = 40$ mm. Additionally, the jet width of the supersonic nozzles was higher compared to the straight nozzles at all locations.

Our previous study [5] found that ignition in the main chamber heavily depends on the mixing process between the hot jet and the cold ultra-lean H_2/air mixture in the ambient. Therefore, it is worth investigating the vorticity field of the hot jet. Vortices help in molecular mixing of chemical species and transfer of momentum and energy. Motivated by this, the vorticity distribution was plotted in Fig. 4.22 at four different axial locations, $x = 10$, 20, 30, and 40 mm, respectively, prior to main chamber ignition at the lean flammability limit of the corresponding nozzles. Vorticity, $\nabla \times \vec{u}$, can be written as [36]

$$\omega_i = \epsilon_{ijk} \frac{\partial u_k}{\partial x_j} \tag{4.18}$$

where ϵ_{ijk} is the alternating tensor. A higher value of vorticity implies a stronger mixing of the hot jet with the cold unburned main chamber fuel/air mixture. Figure 4.22 shows that the vorticity generated by the supersonic nozzles is higher compared to the straight nozzles.

Another interesting fact is that the vorticity distributions were wider for supersonic nozzles. Unlike the straight nozzles, the supersonic nozzles (nozzle 4 and nozzle 5), which were able to ignite leaner H_2/air mixtures at $\phi = 0.22$ and 0.23, respectively, had a wider vorticity distribution between 30 mm $< x <$ 40 mm.

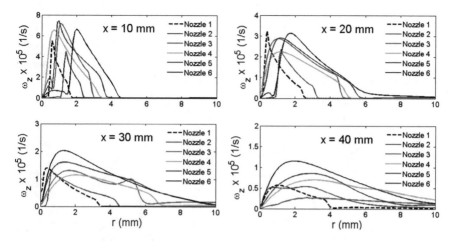

Fig. 4.22 Vorticity distributions at different axial locations, $x = 10, 20, 30,$ and 40 mm, just before ignition happened in the main chamber at lean flammability limit of corresponding nozzles as shown in Table 4.1

Species Concentration

The ignition probability highly depends on the local species concentration as well as the temperature distribution. Since the lean flammability limit was different for all six nozzles, the unburned H_2 concentration in the main chamber mixture was varied as well. For comparison, we used a normalized H_2 mass fraction defined as

$$\zeta = \frac{Y_{H_2}}{Y_{H_2|Ub}} \tag{4.19}$$

Here Y_{H_2} is the mass fraction of H_2 in the mixture, and $Y_{H_2|Ub}$ is the mass fraction of the unburned H_2 at $t = 0$ in the main chamber. This normalized ratio indicates the degree of mixing between the hot combustion products and the cold unburned ambient mixture in the form of species concentration. Figure 4.23 shows the normalized H_2 mass fraction, ζ, just prior to main chamber ignition for all the nozzles at their, respectively, lean flammability limit.

Since the jet width varies depending on the nozzle, we have fixed the origin of the radial coordinate at $\zeta = 0.5$. It is evident from Fig. 4.23 that at $x = 10$ mm, the diffusion of H_2 was limited to the mixing layer at $-0.3 < r < 0.3$ mm. As the jets penetrated further into the main chamber, more of the unburned H_2 started to diffuse into the hot jet. However, the diffusion of the unburned H_2 for straight nozzles was slower compared to supersonic nozzles. This is evident from the normalized species concentration profiles at $x = 40$ mm. At $x = 40$ mm, the width of the diffusion zone for the supersonic nozzles was $-2 < r < 2$ mm, whereas for the straight nozzles it was $-1.2 < r < 1.2$ mm.

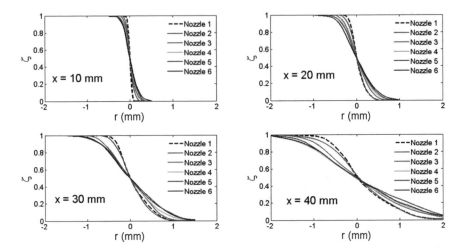

Fig. 4.23 Normalized H_2 species mass fraction, ζ, distributions at different axial locations, $x = 10$, 20, 30, and 40 mm, just before ignition took place in the main chamber at the lean flammability limit of corresponding nozzles as shown in Table 4.1

Temperature

Figure 4.24 shows the spatial distribution of the static temperature of the jets just before ignition occurred in the main chamber at their respective lean flammability limit. As the hot jet penetrated into the main chamber, it cooled down due to mixing with the cold ambient gases. As a result, for the two straight nozzles, the jet temperature decreased monotonically from the exit to the downstream locations. For supersonic nozzles, however, the static temperature fluctuated due to the presence of shock diamonds. The static temperature increased after each shock. Finally, a higher temperature zone after the final strong shock was observed for both convergent and C-D nozzles. However, the length and the starting position of this high-temperature zone differed for different nozzles. Table 4.4 shows the starting position and the length of this high-temperature zone. It is interesting to notice that even though the starting position of the high-temperature zone was different for different nozzles, nozzle 4 and 5 had a wider high-temperature zone when compared to the other supersonic nozzles.

The computed normalized instantaneous jet centerline temperature profiles, T/T_{max}, are shown in Fig. 4.25. Inside the potential core of the jet, the centerline temperature of the two straight nozzles decreased only slightly. However, there was a sharp drop in temperature just outside of the potential core. Finally, the temperature far downstream approached the ambient temperature. For both converging and C-D nozzles, the centerline temperature fluctuated near the jet exit due to the presence of shocks and expansion fans. After the final strong shock, the centerline temperature rise occurred for all four supersonic nozzles. The location and size of the high-temperature zone are shown in Table 4.4. After the final shock, the supersonic

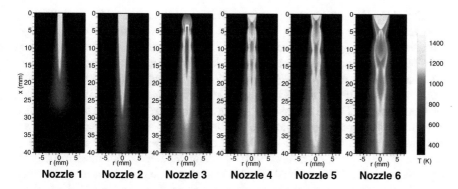

Fig. 4.24 Spatial distribution of temperature profiles at an instance just before ignition initiated in the main chamber at the lean flammability limit of corresponding nozzles as shown in Table 4.1

Table 4.4 Starting position and the length of the high-temperature zone for different nozzles

Nozzle #	Starting position, mm	Length, mm
Nozzle 1, straight, $d = 1.5$ mm	–	–
Nozzle 2, straight, $d = 3$ mm	–	–
Nozzle 3, convergent, $A_i/A_e = 4$	13	20
Nozzle 4, C-D, area ratio $= 4$	15	16
Nozzle 5, C-D, area ratio $= 9$	22	17
Nozzle 6, C-D, area ratio $= 16$	24	7.5

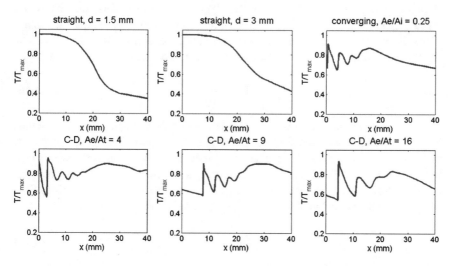

Fig. 4.25 Computed instantaneous temperature profiles at the jet centerline for different types of nozzle designs just before ignition occurred in the main chamber at the lean flammability limit of corresponding nozzles as shown in Table 4.1

nozzles showed normalized temperature, $T/T_{\max}$, in the range of $0.7 - 0.85$ at the downstream location 13 mm $< x <$ 24 mm, whereas at the same location straight nozzles had $T/T_{\max} = 0.5 - 0.6$. Thus, the temperature of the straight jets was 26–30% lower compared to supersonic jets inside the downstream region of 13 mm $< x <$ 24 mm. Although the location and length of this high-temperature zone varied for different nozzles, the much interesting fact lies in the ignition pattern of these nozzles. For all supersonic nozzles, ignition in the main chamber started from this high-temperature zone. Even though nozzle 6 showed a high-temperature zone near $x = 40$ mm and nozzle 3 showed a high-temperature zone in the range of 20 mm $< x <$ 30 mm, both failed to ignite main chamber H_2/air mixtures with an equivalence ratio of ϕ less than 0.29. This indicates, even though the jet temperature is an important factor, however, temperature alone does not control the ignition probability. This will be addressed in the next section.

4.4.2.4 Damköhler Number

As soon as the turbulent hot jet enters the main chamber, it starts mixing with the cold, unburned premixed H_2/air in the main chamber. The mixing layer contains many small eddies. As turbulent eddies dissipate energy to the cold unburned ambient gases, the temperature of the hot jet drops as it goes further downstream. The competition between turbulent mixing and chemical reaction, characterized by the Damköhler number, has a deterministic effect on the ignition outcome. Motivated by this, we plotted the Damköhler number profiles of the jet just prior to main chamber ignition for six different nozzles. This will help us understand why nozzle 4 and 5 can extend the lean flammability limit of the mixture in the main chamber.

As discussed in Chap. 2, the Damköhler number, Da, is defined as the ratio of the characteristic flow timescale τ_F to the characteristic chemical reaction timescale, τ_C.

$$Da = \frac{\tau_F}{\tau_C} \tag{4.20}$$

Here the characteristic flow timescale is the turbulent mixing timescale, which largely depends on the turbulent mean and fluctuation velocities (or the Reynolds number). The characteristic chemical timescale is the ignition timescale, which mainly depends on the chemistry of the reactions and the temperature at which the reactions take place. The Damköhler number can further be written in terms of the fuel/air thermophysical properties and flow field information as [37]:

$$Da = \frac{s_L l}{u' l_f} \tag{4.21}$$

where s_L is the laminar flame speed, l is the integral length scale, u' is the fluctuating component of the velocity, and l_f is the flame thickness. Laminar burning speed, s_L, was calculated using the chemistry model for H_2/air as described in Sect. 4.3.4. Laminar flame thickness, l_f, was calculated using the PREMIX module of

ChemkinPro [38]. The spatial distribution of $u'$and l was calculated using the following relations [37]:

$$u' = \bar{u}I \qquad (4.22)$$

$$l = \int_0^\infty f(r)dr; f(r) = 1 - \frac{3}{4}\frac{C}{k}(\epsilon r)^{\frac{2}{3}} \qquad (4.23)$$

where $\bar{u}$ is the mean velocity of the flow field, I is the turbulent intensity, ϵ is the turbulent dissipation rate, k is the turbulent kinetic energy, r is the spatial coordinate, and C is a universal constant called the Kolmogorov constant.

Figure 4.26 shows the Damköhler number profiles just prior to main chamber ignition for all six nozzles at their respective lean flammability limit (the limit is reported in Table 4.1). The highest Damköhler numbers occurred in the mixing layer between the hot jet and the cold ambient. The thickness and the length of the highest Damköhler number region changed with nozzle type. Since the six cases shown were all at their respective lean flammability limit, we defined a limiting Damköhler number, $Da_{critical}$, below which main chamber ignition would be impossible to occur. We found this limiting Damköhler number to be around 11 based on the local flow conditions. Note our previous study [5] suggested that for successful ignition of H_2/air in the main chamber, the critical Damköhler number should be 40. However, the calculations in [5] were based on the exit jet conditions. It did not consider the local flow properties. That was why we called it the global Damköhler number. In the present study, we could obtain local information such as velocity, fluctuations, turbulent intensity, species mass fraction, and temperature. Thus, the Damköhler number profiles shown in Fig. 4.26 better reflect the interaction between turbulence and chemistry locally. Figure 4.26 clearly explains why the converging and C-D nozzles performed better than the straight nozzles, from the perspective of Damköhler numbers. It also helps to explain where along the jet ignition is likely to occur. For example, for nozzle 2 the chances are high at the jet surface between 25 mm $< x <$ 35 mm, and this location is consistent with the experimental

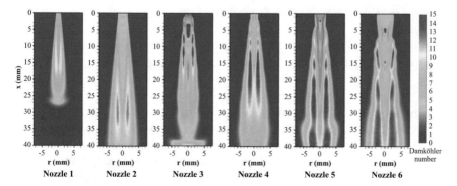

Fig. 4.26 Spatial distribution of Damköhler number just before ignition occurred in the main chamber for different nozzles at their respective lean flammability limit as reported in Table 4.1

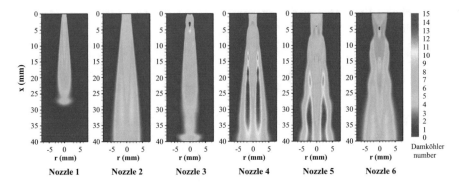

Fig. 4.27 Spatial distribution of the Damköhler number just before ignition occurred in the main chamber for all six nozzles at $\phi = 0.23$

observations. For nozzle 4 the ignition probability is higher between 10 mm $< x <$ 30 mm, where the Damköhler number values are the highest.

Among the three C-D nozzles, nozzles 4 and 5 with an area ratio of 4 and 9, respectively, performed better than the one with an area ratio of 16. The former two resulted in a lean flammability limit, ϕ_{limit}, of 0.22 and 0.23, respectively, while the latter exhibited a flammability limit, ϕ_{limit}, of 0.29. To further understand the effect of area ratio, we filtered the effect of flammability limits and plotted the spatial distribution of the Damköhler number for all three nozzles at the same equivalence ratio, $\phi = 0.23$, as shown in Fig. 4.27. It is clear that nozzle 4 and 5 produced higher Damköhler numbers in the mixing layer than nozzle 6. This indicates that a supersonic jet with higher Mach numbers reduces ignition probability. Note all three C-D nozzles produced a high-temperature zone downstream of the nozzle exit as reported in Table 4.4, which would result in a higher likelihood of ignition. Because nozzle 6 had the largest aspect ratio among the three, the velocity at the nozzle exit and downstream was greater than that of the other two nozzles. As a result, turbulent mixing was too fast between the jet produced by nozzle 6 and the cold ambient. Thus, it failed to extend the lean flammability limit.

4.4.2.5 Burning Time of the Main Chamber Mixture

Previous sections explain the mechanism why supersonic hot jet can extend the lean flammability limit of the H_2/air mixture in the main chamber. In this section, we compared the combustion efficiency of these six nozzles by examining the total burning time required to consume all the reactants in the main chamber. Figure 4.28 shows the overall burning time of six nozzles at different main chamber equivalence ratios. These burning times were measured experimentally and present an overall comparison between different nozzles from the combustion efficiency viewpoint. Since turbulent flame speed, s_T, depends on the fluctuating component of velocity, $u^{'}$ [37]:

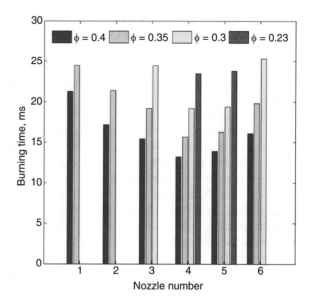

Fig. 4.28 Comparison of the experimentally measured burning time for different nozzles

$$\frac{s_T}{s_L} = \sqrt{\frac{u'l}{s_L l_f}} \tag{4.24}$$

Figure 4.28 shows that the turbulence intensity of different nozzles varies and in turn affect the turbulent flame propagation speed. Since supersonic jets increase turbulence and mixing, it may facilitate the burning speed in the main chamber. Figure 4.28 shows that the supersonic nozzles exhibited a lower burning time as compared to the straight nozzles. Moreover, nozzles 3 and 4 had the lowest burning time compared to the other nozzles. As the fuel/air mixture became leaner, the burning time increased since the flame propagation speed decreased. For example, the burning time of nozzle 4 at $\phi = 0.4$ is 13.2 ms and at $\phi = 0.23$ is 23.5 ms. On the other hand, the burning time for nozzle 1 at $\phi = 0.4$ is 21.3 ms.

4.5 Conclusions

Present work investigated the ignition characteristics of ultra-lean H_2/air mixtures using a supersonic hot jet generated by pre-chamber combustion both experimentally and numerically. The first half of the chapter discusses the experimental findings. High-speed schlieren, OH* chemiluminescence, and IR imaging were applied to visualize the jet penetration and ignition processes. However, experimental results alone could not explain the ignition behavior and flammability limits. This forces us to consider the problem numerically. The second half of the chapter explores the numerical results of the hot turbulent jet.

A vital finding from experiments is the extension of lean limit, ϕ_{limit}, and lower ignition delay of the lean H_2/air mixture in the main chamber by using a supersonic nozzle than a straight nozzle. Ignition in the main chamber was achieved for $\phi = 0.22$ using a supersonic nozzle. Simultaneous schlieren and OH* chemiluminescence results show ignition initiates from the side surface of the hot jet. Due to the presence of shock structures at the exit of a supersonic jet, supersonic jet exit temperature is higher than straight nozzles. The increase in the static temperature behind the shocks thus escalates ignition probability, which is the main reason that the lean limit can be further reduced. Moreover, converging and C-D nozzles create a high-temperature zone downstream of shocks responsible for initiation of ignition. This is a key piece of information. This could help us better control the ignition location and ignition delays and design a better pre-chamber for lean combustion. However, the experimental results do not explain why the ignition started from the high-temperature zone or why only nozzles with area ratio 4 and 9 were successful to extend the lean flammability limit of H_2/air.

To understand the problem completely, the ignition characteristics of ultra-lean H_2/air mixtures by a hot supersonic jet was examined numerically. The transient flame propagation process in the pre-chamber filled with stoichiometric premixed H_2/air was examined. Just after ignition in the pre-chamber, a hemispherical flame started propagating outwardly at a constant speed – laminar burning speed. As the flame surface area increased, the flame started to accelerate until it touched the wall. Then the flame stopped accelerating due to a sudden reduction in the flame surface area. Afterward, the flame front changed into a tulip shape due to the complex interaction between the reflected pressure waves and the flame front. The tulip shape caused the flame to accelerate again until it reached the connecting nozzle inlet.

The mechanism why the supersonic jets can extend the lean flammability limit of H_2/air mixture was explained. Due to higher velocity and vorticity, the supersonic jets could mix with the cold unburned H_2/air more efficiently than subsonic nozzles. Simultaneously, the static temperature of the supersonic jets increased after each shock, and after the final strong shock, the temperature rise was significant. Main chamber ignition was initiated from this high-temperature region. These two phenomena together raised the possibility of ultra-lean ignition using supersonic jets over subsonic jets. The ignition criteria could be characterized by the Damköhler number. The critical Damköhler number was found to be 11 for H_2/air based on local flow and chemical timescales. The probability of ignition is nearly zero below this critical Damköhler number.

Lastly, not all the supersonic jets could extend the lean flammability limit of H_2/air. The convergent nozzle and C-D nozzle with an area ratio of 16 failed to do so. The reason is that excessive turbulence from these supersonic nozzles resulted in rapid mixing and lower Damköhler numbers.

References

1. Djebaili, N., et al.: Ignition of a combustible mixture by a hot unsteady gas jet. Combust. Sci. Technol. **104**(4–6), 273–285 (1995)
2. Boretti, A.A.: Modelling auto ignition of hydrogen in a jet ignition pre-chamber. Int. J. Hydrog. Energy. **35**(8), 3881–3890 (2010)
3. Boretti, A.A., Watson, H.C.: The lean burn direct injection jet ignition gas engine. Int. J. Hydrog. Energy. **34**(18), 7835–7841 (2009)
4. Chiera, D., Riley, M., Hampson, G.J.: Mechanism for high-velocity turbulent jet combustion from passive prechamber spark plug. In: Proceedings of the ASME 2012 Internal Combustion Engine Division Fall Technical Conference, Vancouver, BC, Canada (2012)
5. Biswas, S., et al.: On ignition mechanisms of premixed CH4/air and H2/air using a hot turbulent jet generated by pre-chamber combustion. Appl. Therm. Eng. **106**, 925–937 (2016)
6. Biswas, S., Qiao, L.: Prechamber hot jet ignition of ultra-lean H2/air mixtures: effect of supersonic jets and combustion instability. SAE Int. J. Engines. **9**(3), 1584–1592 (2016)
7. ANSYS: ANSYS Fluent Academic Research, Release 15.0 (2015)
8. Launder, B.E., Reece, G.J., Rodi, W.: Progress in the development of a Reynolds-stress turbulent closure. J. Fluid Mech. **68**(3), 537–566 (1975)
9. Jones, W.P., Launder, B.E.: The prediction of laminarization with a two-equation model of turbulence. Int. J. Heat Mass Transf. **15**(2), 301–314 (1972)
10. Launder, B.E., Sharma, B.I.: Application of the energy dissipation model of turbulence to the calculation of flow near a spinning disc. Lett. Heat Mass Transf. **1**(2), 131–138 (1974)
11. Menter, F.R.: *Zonal Two Equation k-ω Turbulence Models for Aerodynamic Flows*. AIAA Paper 93-2906, NASA-TM-111629 (1993)
12. Daly, B.J., Harlow, F.H.: Transport equations in turbulence. Phys. Fluids. **13**(11), 2634–2649 (1970)
13. Speziale, C.G., Sarkar, S., Gatski, T.B.: Modeling the pressure-strain correlation of turbulence – an invariant dynamical systems approach. J. Fluid Mech. **227**, 245–272 (1991)
14. Connaire, M.O., et al.: A comprehensive modeling study of hydrogen oxidation. In. J. Chem. Kinet. **36**(11), 603–622 (2004)
15. Magnussen, B.F.: On the structure of turbulence and a generalized eddy dissipation concept for chemical reaction in turbulent flow. In: Nineteenth AIAA Meeting, St. Louis (1981)
16. Iijima, T., Takeno, T.: Effects of temperature and pressure on burning velocity. Combust. Flame. **65**(1), 35–43 (1986)
17. Patankar, S.V.: Numerical Heat Transfer and Fluid Flow. Hemisphere Series on Computational Methods in Mechanics and Thermal Science. CRC Press (1980) https://www.crcpress.com/Numerical-Heat-Transfer-and-Fluid-Flow/Patankar/p/book/9780891165224
18. VanLeer, B.: Towards the ultimate conservative difference scheme, V. A second order sequel to Godunov's method. J. Comput. Phys. **32**, 101–136 (1979)
19. Matalon, M., Daou, J.: Influence of conductive heat-losses on the propagation of premixed flames in channels. Combust. Flame. **128**, 321–339 (2002)
20. Matalon, M.: Flame dynamics. Proc. Combust. Inst. **32**(1), 57–82 (2009)
21. Bychkov, V., Fru, G., Petchenko, A.: Flame acceleration in the early stages of burning in tubes. Combust. Flame. **150**(4), 263–276 (2007)
22. Gonzalez, M., Borghi, R., Saouab, A.: Interaction of a flame front with its self-generated flow in an enclosure; the tulip flame phenomenon. Combust. Flame. **88**(2), 201–220 (1992)
23. Dunn-Rankin, D., Sawyer, R.F.: Tulip flames: changes in shape of premixed flames propagating in closed tubes. Exp. Fluids. **24**(2), 130–140 (1998)
24. Bychkov, V.V., Liberman, M.A.: Dynamics and stability of premixed flames. Phys. Rep. **325**(4), 115–237 (2000)
25. Xiao, H., et al.: An experimental study of distorted tulip flame formation in a closed duct. Combust. Flame. **160**(9), 1725–1728 (2013)

26. Xiao, H., et al.: Dynamics of premixed hydrogen/air flame in a closed combustion vessel. Int. J. Hydrog. Energy. **38**(29), 12856–12864 (2013)
27. Xiao, H., et al.: Experimental study on the behaviors and shape changes of premixed hydrogen–air flames propagating in horizontal duct. Int. J. Hydrog. Energy. **36**(10), 6325–6336 (2011)
28. Ponizy, B., Claverie, A., Veyssière, B.: Tulip flame – the mechanism of flame front inversion. Combust. Flame. **161**(12), 3051–3062 (2014)
29. Markstein, G.H.: Nonsteady Flame Propagation AGARDograph, vol. 75. Elsevier Science, Burlington (2014)
30. Marra, F.S., Continillo, G.: Numerical study of premixed laminar flame propagation in a closed tube with a full navier-stokes approach. Int. Symp. Combust. **26**(1), 907–913 (1996)
31. Lovely, D., Haimes, R.: Shock detection from computational fluid dynamics results. In: 14th Computational Fluid Dynamics Conference. The American Institute of Aeronautics and Astronautics (AIAA), Norfolk (1999) https://doi.org/10.2514/6.1999-3285; https://arc.aiaa.org/doi/abs/10.2514/6.1999-3285
32. Anderson, J.D.: Fundamentals of Aerodynamics. Anderson Series, vol. xxiii, 5th edn, p. 1106. McGraw-Hill, New York (2011)
33. Wishart, D.P., Krothapalli, A.: The structure of a heated supersonic jet operating at design and off-design conditions, ProQuest Dissertations and Theses, p. 9526501. The Florida State University, ProQuest Dissertations Publishing (1995)
34. Barlow, R., et al.: Structure of a supersonic reacting jet. In: 29th Aerospace Sciences Meeting, Reno, Nevada (1991)
35. Gutmark, E., Schadow, K.C., Bicker, C.J.: Near acoustic field and shock structure of rectangular supersonic jets. AIAA J. **28**(7), 1163–1170 (1990)
36. Kundu, P.K., Cohen, I.M., Dowling, D.R.: Fluid Mechanics, vol. xxvi, 5th edn, p. 891. Academic Press, Waltham (2012)
37. Peters, N.: Turbulent Combustion. Cambridge Monographs on Mechanics, vol. xvi, p. 304. Cambridge University Press, Cambridge, UK/New York (2000)
38. Reaction Design: Reaction Workbench 15131 San Diego (2013) http://www.reactiondesign.com/support/help/help_usage_and_support/how-to-citeproducts/

Chapter 5
Combustion Instability at Lean Limit

Contents

5.1   Introduction

In recent years gas engine manufacturers have faced stringent emission regulations on oxides of nitrogen (NOx) and unburned hydrocarbons (UHC) [1, 2]. Operating internal combustion engines at ultra-lean conditions can reduce NO_x emissions and also improve thermal efficiency [3, 4]. An approach that can potentially solve the challenge of igniting ultra-lean mixtures is to use a reacting/reacted hot turbulent jet to ignite the ultra-lean mixture instead of a conventional electric spark [5–10]. The hot turbulent jet is produced by burning a small amount of stoichiometric or near-stoichiometric fuel/air mixture in a small volume separated from the main combustion chamber called the pre-chamber. The higher pressure resulting from pre-chamber combustion pushes combustion products into the main combustion

© Springer International Publishing AG, part of Springer Nature 2018   101
S. Biswas, *Physics of Turbulent Jet Ignition*, Springer Theses,
https://doi.org/10.1007/978-3-319-76243-2_5

chamber in the form of a hot reacting/reacted turbulent jet, which then ignites the ultra-lean mixture in the main combustion chamber. Compared to conventional spark ignition, the hot turbulent jet has a much larger surface area containing numerous ignition kernels over which ignition can occur. Hot jet ignition has the potential to enable the combustion system to operate near the fuel's lean flammability limit, leading to ultralow emissions.

However, ignition and combustion of ultra-lean fuel/air mixtures have many challenges. The first challenge regards reliable ignition at ultra-lean conditions. For example, due to poor ignition, misfires can occur in the engine. Such misfires can lead to cycle-to-cycle variability, rough operation, increase in unburned hydrocarbon emissions, and reduction in efficiency – none of which are desirable [11, 12]. The second challenge deals with the unstable combustion dynamics at the lean limit. As the operating condition moves toward lean flammability limit, the flame becomes weaker, and a further response by growing acoustic disturbance field leads to thermoacoustic oscillations. Characteristics of thermoacoustic instability include large pressure oscillations, oscillating flame propagation, and reduction in flame speed, i.e., reduction in stable flame propagation limit in general. This has an adverse effect on combustion efficiency and also threatens the structural life of combustor [13–15].

Combustion instability is a highly complex phenomenon caused by coupling between pressure and heat release fluctuations in a combustion chamber. While a large number of literature on thermoacoustic instability exist for solid rockets [16–20], liquid rockets [21–26], and gas turbines [27–33], very few are available for pre-chamber jet ignition combustion systems, such as heavy-duty, lean-burn jet ignited or stratified natural gas engines. In our previous studies [34, 35], we found that supersonic jets not only shortened the ignition delay but also extended the lean flammability limit of H_2/air mixture in the main chamber, as compared to the jets produced by straight nozzles. For example, the lean limit achieved using straight nozzles was found to be $\phi = 0.31$. Supersonic converging-diverging nozzles, however, can extend this limit to $\phi = 0.22$. Nevertheless, thermoacoustic instability became severe at the ultra-lean conditions $0.22 < \phi < 0.3$. The existence of combustion instability was inferred from visual observations of oscillating flame front via high-speed schlieren and OH* chemiluminescence imaging as well as unsteady pressure oscillations. Strain gauge and accelerometer measurements attached to the combustor show increase in structure vibration during ultra-lean operations. Controlling such instability, active or passive, requires adequate knowledge about different types of instability modes, perturbation energy, and frequencies associated with the instability.

Motivated by the above, this chapter investigates the characteristics of thermoacoustic instability that arises during the pre-chamber hot jet ignition of ultra-lean premixed H_2/air using both experiment and mathematical modeling. The self-excited oscillation of the hot jet that initiated ignition in the main chamber and the effect of equivalence ratio on instability were discussed. Combustion instability was characterized using experimentally determined fast Fourier transform (FFT), Rayleigh criteria, power spectra, bifurcation behavior, and growth and decay rates.

These results were then compared with the modeling results by solving the 1D and 3D linearized Euler equations (LEEs) coupled with combustion response functions (CRFs) in the frequency domain. A phase-resolved measurement of strain rates along the oscillating flame edge was carried out. Finally, the unstable modes and their behavior were identified using dynamic mode decomposition (DMD), and a combustion instability mechanism was proposed.

5.2 Experimental Method

The experimental method is discussed in detail in [34, 35] and will be briefly introduced here. The experimental setup is schematically shown in Fig. 5.1a and b [28]. A small volume, 100 cc cylindrical stainless steel (SS316) pre-chamber was attached to the rectangular ($17'' \times 6'' \times 6''$) carbon steel (C-1144) main chamber. The main chamber to pre-chamber volume ratio was 100:1. A stainless steel orifice plate with a constant nozzle length (10 mm) and various nozzle designs (reported in Table 5.1) separated both chambers. Combustion instability characteristics of H_2/air for six different nozzle designs were investigated.

A thin, 25 μm thick aluminum diaphragm separated both chambers with dissimilar equivalence ratios from mixing. In the present study, all tests were conducted at room temperature 300 K. The stoichiometric fuel/air mixture in the pre-chamber was ignited by an electric spark created from a 0–40 kV capacitor discharge ignition (CDI) system. A Bosch iridium spark plug with ultrathin iridium center electrode was attached at the top of the pre-chamber. The transient pressures of both chambers were measured using high-resolution ~ 10 kHz Kulite (XTEL-190) pressure transducers and were recorded using NI-9237 signal conditioning and pressure acquisition module by LabVIEW software. A $1''$ thick polymer insulation jacket was wrapped around the pre-chamber and the main chamber to minimize heat loss. The amount of fuel (industrial grade H_2, 99.98% pure) and air were calculated using partial pressure method and were introduced using two separate ports into the constant-volume main chamber. Unlike the main chamber where fuel and air were mixed in the chamber itself, fuel/air in the pre-chamber was premixed in a small stainless steel mixing chamber (2.5 cm diameter, 10 cm long) prior going into pre-chamber.

5.3 Mathematical Modeling

5.3.1 Modeling of Combustion Instability

Mathematical modeling was performed to determine various thermoacoustic instability modes and to gain insights into the origin and growth of the instabilities. First, a reduced order, 1D linearized Euler equations (LEEs) [36, 37] coupled with

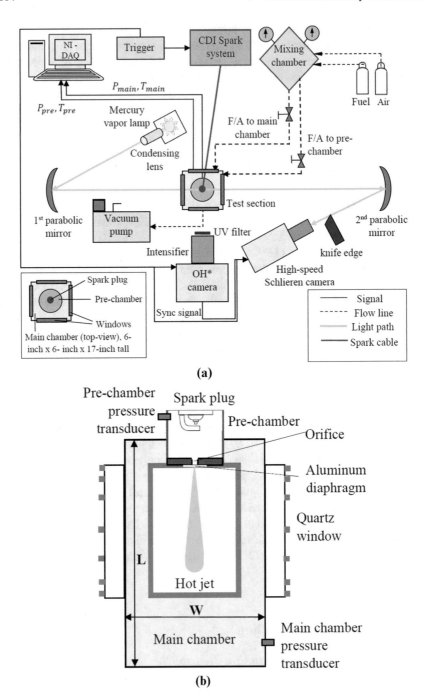

Fig. 5.1 Schematic of (**a**) experimental setup, (**b**) pre-chamber and main chamber assembly [34]

Table 5.1 Dimensions of
the nozzles used in the
experiment [35]

d_{throat}, mm	d_{exit}, mm	A_e/A_t
1.5	1.5	–
3	3	–
–	1.5	–
1.5	3	1
1.5	4.5	9
1.5	7.5	16

combustion response functions (CRF) were solved in the frequency domain to predict the fundamental modes of combustion instabilities in the dual-chamber combustor system. Then 3D linearized Euler equations coupled with advanced combustion response functions (ACRF) were solved based on the exact combustor geometry using COMSOL [38]. The governing equations for linearized Euler equations (LEEs), wave equations, and its implementation in the frequency domain are discussed below. The conservation equations (mass, momentum, and energy) for inviscid flow are

$$\frac{\partial \rho}{\partial t} + \frac{\partial(\rho u)}{\partial x} = 0 \tag{5.1}$$

$$\rho \frac{\partial u}{\partial t} + \rho u \frac{\partial u}{\partial x} + \frac{\partial p}{\partial x} = 0 \tag{5.2}$$

$$\rho \frac{\partial e}{\partial t} + \rho u \frac{\partial e}{\partial x} = -p \frac{\partial u}{\partial x} + q \tag{5.3}$$

where t, x, ρ, u, p, e, and q are time, space, density, velocity, pressure, specific internal energy, and heat release, respectively. The equation of state can be written as

$$p = \rho R T \tag{5.4}$$

$$e = C_v T \tag{5.5}$$

From Eq. (5.5), expressing specific heat at constant volume, C_v in terms of specific heat ratio, γ and substituting into Eq. (5.3), we get

$$e = C_v \frac{p}{R} = \frac{p}{\gamma - 1} \tag{5.6}$$

$$\frac{\partial p}{\partial t} + u \frac{\partial p}{\partial x} + \gamma p \frac{\partial u}{\partial x} = (\gamma - 1)q \tag{5.7}$$

Perturbing the flow by a small amount over mean quantities, we can express pressure, velocity, density, and heat release as a sum of their mean and fluctuating components.

$$p = \bar{p} + p' \tag{5.8}$$

$$u = \bar{u} + u' \tag{5.9}$$

$$\rho = \bar{\rho} + \rho' \tag{5.10}$$

$$q = \bar{q} + q' \tag{5.11}$$

where indicates mean quantities and $'$ indicates fluctuating quantities. Substituting Eqs. (5.8, 5.9, 5.10, and 5.11) into Eq. (5.1, 5.2, and 5.7) and neglecting higher-order terms produce

$$\frac{\partial \rho'}{\partial t} + \frac{\partial}{\partial x}\left(\rho'\bar{u} + \bar{\rho}u'\right) = 0 \tag{5.12}$$

$$\frac{\partial u'}{\partial t} + \bar{u}\frac{\partial u'}{\partial x} + u'\frac{\partial \bar{u}}{\partial x} + \frac{\rho'\bar{u}}{\bar{\rho}}\frac{\partial \bar{u}}{\partial x} + \frac{1}{\bar{\rho}}\frac{\partial p'}{\partial x} = 0 \tag{5.13}$$

$$\frac{\partial p'}{\partial t} + \bar{u}\frac{\partial p'}{\partial x} + u'\frac{\partial \bar{p}}{\partial x} + \gamma p'\frac{\partial \bar{u}}{\partial x} + \gamma \bar{p}\frac{\partial u'}{\partial x} = (\gamma - 1)q' \tag{5.14}$$

Using the following mathematical relationships,

$$\bar{u}\frac{\partial u'}{\partial x} = \frac{\partial}{\partial x}\bar{u}u' - \frac{u'\left(\partial \bar{u}\right)}{\partial x} \tag{5.15}$$

$$\frac{1}{\bar{\rho}}\frac{\partial p'}{\partial x} = \frac{\partial}{\partial x}\left(\frac{p'}{\bar{\rho}}\right) - p'\frac{\partial}{\partial x}\left(\frac{1}{\bar{\rho}}\right) \tag{5.16}$$

$$\bar{u}\frac{\partial p'}{\partial x} = \frac{\partial}{\partial x}\bar{u}p' - p'\frac{\partial \bar{u}}{\partial x} \tag{5.17}$$

$$\gamma \bar{p}\frac{\partial u'}{\partial x} = \gamma\frac{\partial}{\partial x}\bar{p}u' - \gamma u'\frac{\partial \bar{p}}{\partial x} \tag{5.18}$$

and substituting Eqs. (5.15, 5.16, 5.17, and 5.18) into Eqs. (5.12, 5.13, and 5.14), we get

$$\frac{\partial \rho'}{\partial t} + \frac{\partial}{\partial x}\left(\rho'\bar{u} + \bar{\rho}u'\right) = 0 \tag{5.19}$$

$$\frac{\partial u'}{\partial t} + \frac{\partial}{\partial x}\bar{u}u' - u'\frac{\partial \bar{u}}{\partial x} + u'\frac{\partial \bar{u}}{\partial x} + \frac{\rho'\bar{u}}{\bar{\rho}}\frac{\partial \bar{u}}{\partial x} + \frac{\partial}{\partial x}\left(\frac{p'}{\bar{\rho}}\right) - p'\frac{\partial}{\partial x}\left(\frac{1}{\bar{\rho}}\right) = 0 \tag{5.20}$$

$$\frac{\partial p'}{\partial t} + \frac{\partial}{\partial x}\bar{u}p' - p'\frac{\partial \bar{u}}{\partial x} + u'\frac{\partial \bar{p}}{\partial x} + \gamma p'\frac{\partial \bar{u}}{\partial x} + \gamma\frac{\partial}{\partial x}\bar{p}u' - \gamma u'\frac{\partial \bar{p}}{\partial x} = (\gamma - 1)q' \tag{5.21}$$

The differential time domain formulation can be converted into the frequency domain, complex eigenvalue formulation, where $^\wedge$ denotes amplitude and w denotes magnitude of complex formulation.

$$p'(x, t) = Re\left(\hat{p}(x).e^{iwt}\right) \tag{5.22}$$

$$u'(x, t) = Re\left(\hat{u}(x).e^{iwt}\right) \tag{5.23}$$

$$\rho'(x, t) = Re\left(\hat{\rho}(x).e^{iwt}\right) \tag{5.24}$$

$$q'(x, t) = Re\left(\hat{q}(x).e^{iwt}\right) \tag{5.25}$$

Substituting Eqs. (5.22, 5.23, 5.24, and 5.25) into Eqs. (5.19, 5.20, and 5.21), we get 1D linearized Euler equations in the frequency domain.

$$i\omega\widehat{\rho} + \frac{\partial}{\partial x}\left(\widehat{\rho}\bar{u} + \bar{\rho}\widehat{u}\right) = 0 \tag{5.26}$$

$$i\omega\widehat{u} + \frac{\partial}{\partial x}\overline{u}\widehat{u} - \widehat{u}\frac{\partial\bar{u}}{\partial x} + \widehat{u}\frac{\partial\bar{u}}{\partial x} + \frac{\bar{\rho}\widehat{u}}{\bar{\rho}}\frac{\partial\bar{u}}{\partial x} + \frac{\partial}{\partial x}\left(\frac{\widehat{p}}{\bar{\rho}}\right) - \widehat{\rho}\frac{\partial}{\partial x}\left(\frac{1}{\bar{\rho}}\right) = 0 \tag{5.27}$$

$$i\omega\widehat{p} + \frac{\partial}{\partial x}\overline{u}\widehat{u} - \widehat{p}\frac{\partial\bar{u}}{\partial x} + \widehat{u}\frac{\partial\bar{p}}{\partial x} + \gamma\bar{p}\frac{\partial\bar{u}}{\partial x} + \gamma\frac{\partial}{\partial x}\overline{p}\widehat{u} - \gamma\widehat{u}\frac{\partial\bar{p}}{\partial x} = (\gamma - 1)\widehat{q} \tag{5.28}$$

The above 1D linearized Euler equations can be generalized in 3D as

$$i\omega\widehat{\rho} + \nabla.\left(\widehat{\rho}\bar{u} + \bar{\rho}\widehat{u}\right) = 0 \tag{5.29}$$

$$i\omega\widehat{u} + \nabla.\left(\widehat{u}.\bar{u} + \frac{\widehat{p}}{\bar{\rho}}I\right) + \frac{\widehat{\rho}}{\bar{\rho}}(\bar{u}.\nabla)\bar{u} - \widehat{p}\nabla\bar{\rho}^{-1} + \left(\nabla\bar{u} - (\nabla.\bar{u})I\right)\widehat{u} = 0 \tag{5.30}$$

$$i\omega\widehat{p} + \nabla.\left(\gamma\bar{p}\widehat{u} + \widehat{p}\bar{u}\right) + (1 - \gamma)(\widehat{u}.\nabla)\bar{p} - (1 - \gamma)(\nabla.\bar{u})\widehat{p} = (\gamma - 1)/\mathbb{Q}_s \tag{5.31}$$

where $\mathbb{Q}_s = \widehat{q}$ W/m^3 is the heat source, u is the velocity field, $\widehat{u}$ denotes complex potential, and ω is the complex argument. Neglecting mean flow in linearized Euler equations gives wave equation. The 1D wave equation can be expressed as

$$\frac{1}{\bar{c}^2}\frac{\partial^2 p'}{\partial t^2} - \nabla^2 p' = \frac{\gamma - 1}{\bar{c}^2}\frac{\partial q'}{\partial t} \tag{5.32}$$

The 3D wave equation in frequency domain can be written as

$$-\frac{1}{\bar{c}^2}\frac{\omega^2\widehat{p}}{\bar{\rho}} - \frac{\nabla^2 p'}{\bar{\rho}} = \frac{\gamma - 1}{\bar{\rho}\bar{c}^2}.i\omega.\widehat{q} \tag{5.33}$$

LEE is a low-order computational method and is often used as an alternative to LES [39] simulations due to its low computational cost and memory requirements. LEE is the system of equations obtained by perturbing the basic conservation equations of mass, momentum, and energy and linearizing them so that they contain only first-order terms. These linearized equations can either be in the time domain or in the frequency domain. In the current study, these equations were solved in the frequency domain. LEE solves for eigenmodes and helps to understand whether the mode is stable or unstable. In addition, the solution also provides information about the "growth rate" of the instability. In general, the lower-order models such as wave equation and LEE are solved in the frequency domain, whereas the higher-order models such as LES are solved in the time domain. It should be noted that the mean flow field is included in the system of equations. Wave equation will be used in cases where there is no mean flow, and LEE will be used when mean flow can no longer be considered negligible.

5.3.2   *Combustion Response Models*

Two different combustion response models – Crocco's n-τ pressure lag model and Dowling's velocity lag model – were implemented in the present study.

5.3.2.1   **Model 1: Crocco's** $n - \tau$ **Model**

Crocco's $n - \tau$ [14, 40] pressure lag model will be implemented to model the unsteady heat release. According to Crocco's model, unsteady heat release fluctuations are related to the pressure fluctuations with a time delay τ.

$$\frac{q'}{\bar{q}} = n \frac{p'(t - \tau)}{\bar{p}} \tag{5.34}$$

where n is a scaling factor that quantifies the intensity of heat release (interaction index) and τ is a time delay between the fuel injection and ignition. After inclusion of combustion response function, the wave equation becomes,

$$-\frac{1}{\bar{c}^2}\frac{\omega^2 \hat{p}}{\bar{\rho}} - \frac{\nabla^2 p'}{\bar{\rho}} = \frac{\gamma - 1}{\bar{\rho}\bar{c}^2}.i\omega.n.\hat{p}.e^{-i\omega\tau}.\frac{\bar{q}}{\bar{p}} \tag{5.35}$$

The heat source term that appeared in Eq. (6.31) can be expressed as

$$Qe = \hat{p}.n.\bar{q}.e^{-i\omega\tau}/\left(\bar{p}.\forall\right) \tag{5.36}$$

where $\forall$ is the volume of heat release zone. For a premixed flame, it is a thin zone separating burned and unburned mixtures.

5.3.2.2   **Model 2: Velocity Lag Model**

In the velocity lag model, unsteady heat release fluctuations $\grave{q}$ are related to velocity fluctuations upstream of heat release zone by a time delay τ. This model was developed by Dowling [31].

$$\grave{q}(x, t) = \grave{Q}(t).\delta(x - b) \tag{5.37}$$

where

$$\grave{Q}(t) = -\frac{\beta\rho\bar{c}^2\grave{u}_1(t - \tau)}{\gamma - 1} \tag{5.38}$$

Substituting Eq. (6.38) in Eq. (6.39) and converting into frequency domain,

$$\widehat{q}(x) = -\frac{\beta\rho\bar{c}^2\widehat{u}_1 e^{-i\omega\tau}\delta(x-b)}{\gamma-1} \tag{5.39}$$

where $Q'(t)$ is heat release per unit area, $\delta(x-b)$ is $1/length\ of\ heat\ release\ zone$, $\widehat{u}_1$ is the velocity fluctuation upstream of flame, and β is the nondimensional quantity that indicates the intensity of heat release. In this model, heat release is considered to be concentrated in small volume. Using Eq. (6.39), the wave equation can be written as

$$-\frac{1}{\bar{c}^2}\frac{\omega^2\widehat{p}}{\bar{\rho}} - \frac{\nabla^2 p'}{\bar{\rho}} = -i\omega\beta\widehat{u}_1 e^{-i\omega\tau}\delta(x-b) \tag{5.40}$$

The heat source term for linearized Euler equations in Eq. (6.31) will become

$$Qe = -\frac{\beta\widehat{u}_1 e^{-i\omega\tau}\delta(x-b)e^{-i\omega\tau}\rho\bar{c}^2}{(\gamma-1)A} \tag{5.41}$$

where A is the cross-sectional area of heat release zone.

Combustion response models were incorporated in the linearized Euler equation solver code or COMSOL as a heat source term in order to account for combustion. For Crocco's n-τ model, the interaction index which signifies the intensity of heat release and the time delay that indicates the time difference between unsteady heat release and pressure fluctuations were calculated from the experimentally measured OH* chemiluminescence and pressure data. For both the models, heat release is assumed to be concentrated in a very thin region. The combustion response model, Crocco's model or Dowling's model, was combined with the linearized Euler equation solver in MATLAB (1D LEE in-house code) or COMSOL (3D LEE). The obtained solution would lie in the complex domain. A positive imaginary part of the complex number indicates decay, and a negative imaginary part indicates growth of instability.

5.4   Results and Discussion

Combustion instability of premixed H_2/air at the ultra-lean limit was examined using both experiment and modeling. First, the experimental results are presented, including characterization of self-excited oscillations of the hot jet, the effect of equivalence ratio on thermoacoustic instability, instability modes calculated using FFT and power spectra, growth and decay rates, bifurcation behavior, unstable flame dynamics, and strain rate probability density functions (PDFs). The modeling results are used to explain different instability modes and their physical significance. The experimentally determined growth rates were compared with the modeling results. The instability modes and their behavior were identified using dynamic mode decomposition (DMD). Finally, a combustion instability mechanism was proposed for combustion systems utilizing pre-chamber hot turbulent jet ignition.

5.4.1 Self-Excited Oscillations of Hot Jet

It is well known that a jet develops an absolute instability in the potential core region if the jet density to the density of the ambient fluid is below a certain critical value [41–43]. Using the spatiotemporal instability analysis and experimental validation, Monkewitz and Sohn [43] showed that the heated jet could become absolutely unstable when the density ratio $S = \rho_j/\rho_\infty < 0.72$, where ρ_j is the jet density and ρ_∞ is the ambient fluid density. In our experiment, the jet-to-ambient density ratio, S, was below 0.36 for all conditions. Along with the critical density ratio, Monkewitz suggested a range of the Strouhal numbers based on nozzle diameter and exit velocity, $St = fd/U_j$, for absolute instability to happen. The Strouhal numbers related to our hot jet were found to be in the range of 0.25–0.5 which falls within the range proposed by Monkewitz. Thus, the pre-chamber combustion-generated hot turbulent jet was absolutely unstable and carried the initial disturbances to the main chamber combustion system. Even though the hot jet was absolutely unstable at all equivalence ratio conditions, combustion instability was observed only for main chamber mixtures having $\phi < 0.5$.

Since an absolutely unstable jet initiated the ignition in the main chamber, we need to understand the global instability modes of the hot jets before going into the details of combustion instability. To capture the unstable modes of the hot jet, the near-field pressure was measured using a high-speed Kulite pressure transducer. The Fourier transform of the near-field pressure reveals the unstable frequencies of the hot jet as shown in Fig. 5.2.

The pressure spectra shown in Fig. 5.2 shows distinct harmonics of 250 Hz. These self-excited oscillations are nothing but the demonstration of a global instability of the unperturbed steady flow. Due to a sudden change in the density across the mixing layer, the globally unstable steady flow will bifurcate and settle into a new organized regime of highly regular oscillations. This new state is termed a global mode of the underlying steady flow, and its oscillations are tuned to a well-defined frequency, which is 250 Hz and its higher harmonics for our hot jet.

It is important to keep in mind that the hot jet issued from the pre-chamber is a highly transient, variable-property hot. For such jets, even slight changes in the velocity and density profiles influence the instability development [44]. Although the unstable frequencies remain the same for a fixed density ratio, a changing jet velocity will change amplification rate of oscillations, or in other words, the relative amplitude of the unstable frequencies will change.

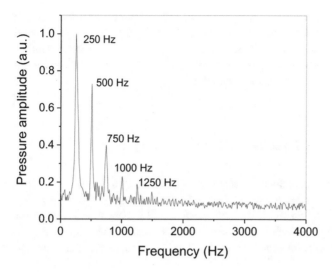

Fig. 5.2 Near-field pressure spectra of hot jet at $x/D = 0.25$ and $r/D = 1$ for $S = \rho_j/\rho_\infty = 0.32$

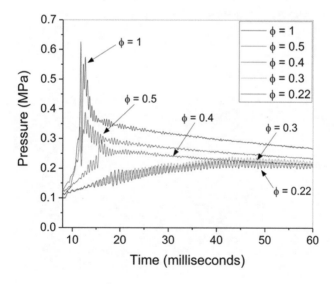

Fig. 5.3 Main chamber pressure data as a function of equivalence ratio

5.4.2 Effect of Pre-chamber Equivalence Ratio

Figure 5.3 shows the main chamber pressure history for various equivalence ratios. Note, for all cases, the pre-chamber mixture was kept at the stoichiometric condition. As the fuel/air mixture becomes leaner, the maximum pressure in the main chamber decreases. Also, the rate of pressure rise decreases as well. As evident from the pressure data in Fig. 5.3, combustion instability becomes predominant at the lean condition. Pressure starts oscillating heavily for $\phi \leq 0.5$. At lean condition, the main chamber burn time is longer compared to the stoichiometric condition. As such, instability gets more time to interact with the chamber acoustics and affects the pressure and in turn flame dynamics.

5.4.3 Fast Fourier Transform (FFT)

Figure 5.4 shows the unsteady pressure histories of the main combustion chamber and the corresponding pressure spectra for various equivalence ratios, ranging from stoichiometric to the lean limit of $\phi = 0.22$. Fast Fourier transform (FFT) was applied to the transient pressure data to determine the frequencies associated with thermoacoustic instability modes. Classical acoustic resonator theory was used to identify the frequencies shown in Fig. 5.4. The dual-chamber system is a closed system with constant volumes. The main combustion chamber can be approximated as 17 inch long, L, and 6 inch wide, W. Hence, the longitudinal and transverse instability modes can be written as [45, 46]

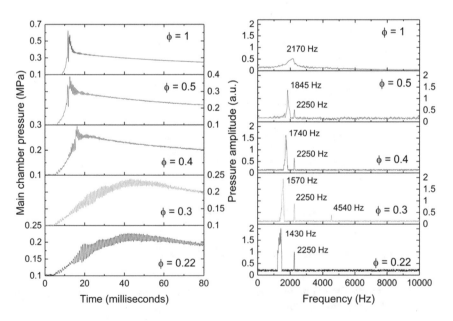

Fig. 5.4 Unsteady main chamber pressure history and corresponding pressure spectra with decreasing ϕ

$$f_1 = \frac{na}{2L} \quad \text{Longitudinal modes} \tag{5.42}$$

$$f_2 = \frac{na}{2W} \quad \text{Transverse modes} \tag{5.43}$$

where $n = 1, 2, 3\ldots$ and $a = \sqrt{\gamma R T_{ad}}$ is the speed of sound. T_{ad} is the constant volume adiabatic flame temperature for the main chamber H_2/air mixture, and the properties such as γ and R were calculated according to the mixture composition in the main chamber using the well-stirred reactor (WSR) model [47]. For the $\phi = 1.0$ case, the first harmonic reveals a single broad frequency peak at about 2170 Hz, which matches well with the calculated frequency of 2155 Hz using Eq. (5.42). As the equivalence ratio decreases, the adiabatic flame temperature T_{ad} drops, which lowers the speed of sound, a. The effect of changes in γ and R is negligible as compared to the flame temperature. As the speed of sound decreases at leaner conditions, decrease in the first harmonic frequency is evident – the peak shifts toward the left of the FFT plot. The measured first harmonic of thermoacoustic instability for $\phi = 0.22$ is 1430 Hz, while it is 2170 Hz for $\phi = 1.0$. At the lean conditions $\phi = 0.22$, 0.3 and 0.5, a second frequency at 2250 Hz was detected. To understand the origin of this frequency, modal analysis of the exact combustor geometry was carried out to find the dynamic response of structure using ANSYS 15.0 [48] modal analysis (FEA module) of the entire combustor. The result revealed that the 2250 Hz mode signifies the transverse deformation of the combustor (which is also the transverse direction of the flame propagation along the jet surface).

5.4.4 Rayleigh Criterion

Thermoacoustic instability arises when an acoustic disturbance grows over time and affects the chamber pressure, which in turn affects the heat release rate and flame propagation speed. A necessary condition for self-excited oscillations is that heat release must supply a net energy to the acoustic disturbance field [46]. Instability grows in time only if oscillating heat release, $q'(t)$, is "in phase," i.e., within $\pm 90°$ with pressure perturbation, $p'(t)$, hence satisfying the "Rayleigh criterion" [46], $\mathbb{R} = \int_0^T p'(t)q'(t)dt > 0$, where $\mathbb{R}$ is the Rayleigh index and T is the period of pressure oscillation. Therefore, a positive Rayleigh index indicates that the unsteady heat release and pressure perturbation are in phase, thus, the instability would grow in time. A negative Rayleigh index denotes that the growth of instability would dampen out since heat release and pressure fluctuation are out of phase.

Figure 5.5 shows the variation of the total OH* chemiluminescence intensity with time over one period of pressure oscillation at the maximum gain condition for $\phi = 0.22$. Since the OH* chemiluminescence intensity is proportional to the flame's heat release, this result agrees with Rayleigh's criterion that, for combustion to be

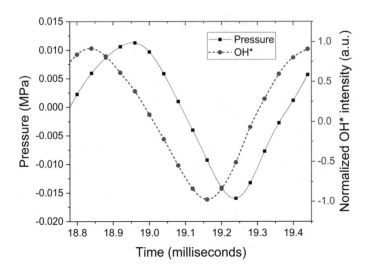

Fig. 5.5 Variation of total OH* intensity and pressure over one period of oscillation at maximum gain for $\phi = 0.22$

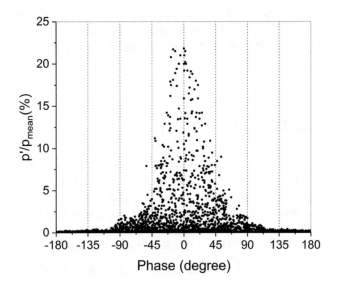

Fig. 5.6 The effect of phase difference between global OH* chemiluminescence and pressure oscillation on the combustion instability pressure amplitude

unstable, the fluctuating heat release must be in phase with the fluctuating pressure. The result in Fig. 5.5 shows that the heat release fluctuation leads the pressure fluctuation by approximately 42.8°.

Figure 5.6 shows the effect of phase on the instability amplitude, p'/p_{mean}. The phase between global OH* chemiluminescence and pressure oscillation was calculated for all the equivalence ratios and plotted with respect to the fluctuating component of pressure. Figure 5.6 shows that for a harmonically oscillating field, the fluctuating heat release adds energy to the local acoustic field when the magnitude of the phase between the pressure and heat release oscillations is less than 90° (i.e., 0 < | Phase| < 90). Conversely, when these oscillations are out of phase (i.e., 90 < | Phase| < 180), the heat addition oscillations damp the acoustic field. Figure 5.6 clearly illustrates that the highest-pressure amplitudes are observed at conditions at which the pressure and heat release are closely in phase. A pressure fluctuation of 5–21.8% over the mean pressure occurs for 0 < |Phase| < 45. The pressure fluctuations decrease below 5% for 45 < |Phase| < 90. Fluctuating pressure amplitude decreases below 1.8% when heat release goes out of phase with pressure oscillations.

5.4.5   Growth and Decay

The hot turbulent transient jet generated from the pre-chamber combustion first accelerates and then decelerates. Our previous studies [34] had shown that the main chamber ignition initiates during the deceleration process of the jet. As soon as the ignition initiates inside the main chamber, the pressure starts rising. For $\phi < 0.5$, during the pressure rise in the main chamber, the pressure starts to oscillate, and the oscillations grow in time. Figure 5.7 shows the pressure fluctuations p' at ultra-lean conditions, $\phi = 0.22$ and 0.3. The amplitude of the disturbance grew in the form of $\sim e^{\alpha t}$, where the growth constant $\alpha > 0$ is the slope of the growth rate. As the main chamber combustion progressed, heat losses to the walls increased as well.

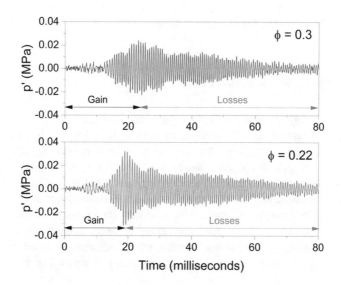

Fig. 5.7 Pressure oscillations due to growing instability for $\phi = 0.22$ and 0.3

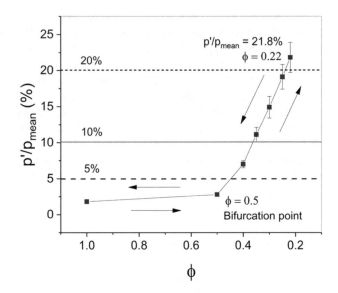

Fig. 5.8 Measured pressure data are illustrating supercritical bifurcation

When main chamber fuel/air completely burned out, the heat release got out of phase with pressure fluctuation, losses was incurred in the system, and pressure perturbation amplitude decayed with time. Like growth, the decay process was nonlinear too, i.e., disturbance amplitude was of the form $(1 - e^{-\beta t})$ where $\beta > 0$ is the decay constant. Unlike rocket or gas turbine instabilities, since there is no continuous mean flow present in the system, both the gain and the loss mechanism last for a brief period of time, in the order of milliseconds. Furthermore, the growth constant α remains positive for 4–15 milliseconds for $\phi < 0.5$.

5.4.6 Supercritical Bifurcation

A bifurcation diagram as a function of the equivalence ratio (the most significant contributing parameter to the instability) serves as an essential element in understanding combustion instability. This diagram also shows whether the point of linear instability (the Hopf bifurcation) is supercritical or subcritical. The nonlinear behavior around the Hopf bifurcation point determines whether it is a subcritical or supercritical bifurcation. Since the gradual change of equivalence ratio triggers the instability, our system shows a supercritical bifurcation. Figure 5.8 shows a smooth, monotonic dependence of the main combustor pressure fluctuations to the mean pressure ratio p'/p_{mean} on the equivalence ratio, indicating a bifurcation in the system. An equivalence ratio of $\phi = 0.5$ separates two regions of fundamentally different dynamics, stable and unstable, respectively, and is referred to as a supercritical bifurcation point [49], below which thermoacoustic combustion instability triggers in. Lastly, near the lean flammability limit of H_2/air ($\phi = 0.22$), the fluctuating pressure p' reaches 21.8% of the mean pressure, p_{mean}. A fluctuating pressure of such intensity can severely damage an engine.

5.4.7 Flame Edge Identification and Strain Rate PDFs

Pressure oscillations coupled with oscillating heat release rate change flame dynamics and burning rate. To understand the effect of pressure perturbations on flame dynamics, strain rates were calculated at four different equidistant phase angles of the oscillating pressure perturbation cycle in the main combustor, labeled from "a" to "d," respectively, as shown in Fig. 5.9a. Then strain rate PDFs were calculated along the oscillating flame edge for these four phases for ultra-lean conditions.

Flame edge was calculated from simultaneous schlieren image velocimetry (SIV) [50] and OH* chemiluminescence images. Given a sufficiently high Reynolds number and adequate refractive flow differences, turbulent eddies can serve as the PIV "particles" in a schlieren or shadowgraph image. Velocity field can be extracted by cross-correlating consecutive high-speed schlieren images. This method is known as SIV and has been demonstrated using a high Reynolds number turbulent helium jet in our previous work [50]. Due to the axisymmetric nature of the flames, the boundaries between burned and unburned regions could be marked using the OH* signal. The strain rates on the flame edge were computed using the SIV data. The two-dimensional stress tensor, $e_{ij} = \frac{1}{2}\left(\frac{\partial U_i}{\partial x_j} + \frac{\partial U_j}{\partial x_i}\right)$, can be written for an axisymmetric geometry in the polar coordinates (r, θ, x) [51] assuming no variation in the θ direction as

$$\kappa = \begin{bmatrix} \dfrac{\partial u_r}{\partial r} & \dfrac{1}{2}\left(\dfrac{\partial u_r}{\partial x} + \dfrac{\partial u_x}{\partial r}\right) \\ \dfrac{1}{2}\left(\dfrac{\partial u_x}{\partial r} + \dfrac{\partial u_r}{\partial x}\right) & \dfrac{\partial u_x}{\partial x} \end{bmatrix} \tag{5.44}$$

This can be written as a symmetric two-dimensional strain tensor containing the normal and shear strain rates that describe fluid element deformation. The final form of the strain rate can be written as

$$\kappa = -n_x n_r \left(\frac{\partial u_x}{\partial r} + \frac{\partial u_r}{\partial x}\right) + \left(1 - n_x^2\right)\frac{\partial u_x}{\partial x} + \left(1 - n_r^2\right)\frac{\partial u_r}{\partial r} \tag{5.45}$$

where the velocity gradients were calculated from the SIV vectors using a fourth-order central difference scheme [52].

High-speed schlieren imaging of flame propagation (top) and the corresponding velocity field (bottom) obtained from SIV (bottom) are shown in Fig. 5.9b and c for $\phi = 0.3$ and $\phi = 0.22$, respectively. It is evident from these figures that the direction

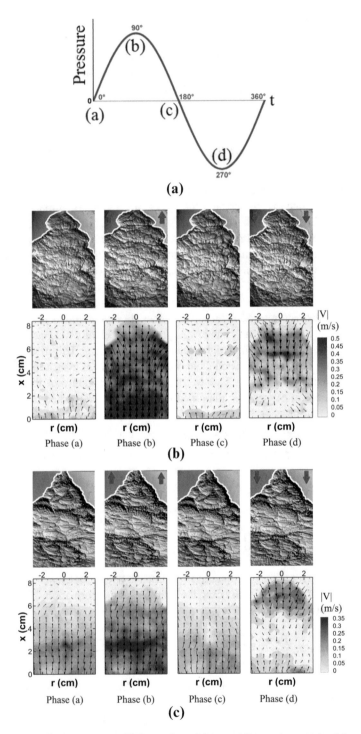

Fig. 5.9 (a) A typical pressure oscillation cycle and four equidistant phase angles (phases "a" to "d"). High-speed schlieren imaging of flame propagation (top) and corresponding velocity field (bottom) for four different phases for (b) $\phi = 0.3$ and (c) $\phi = 0.22$

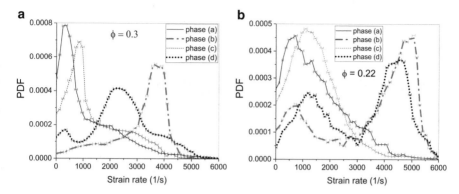

Fig. 5.10 Strain rate PDFs for (**a**) $\phi = 0.3$, most probable strain for phase "a" $= 925$ s^{-1}, phase "b" $= 1110$ s^{-1}, phase "c" $= 3880$ s^{-1}, and phase "d" $= 2790$ s^{-1} and (**b**) $\phi = 0.22$; most probable strain for phase "a" $= 1075$ s^{-1}, phase "b" $= 5040$ s^{-1}, phase "c" $= 1420$ s^{-1}, and phase "d" $= 4390$ s^{-1}

of flame propagation reversed in one pressure perturbation cycle. Global flame surface, $G(\vec{x}, \vec{r}, t)$, moves outward at phase (b) and inward at phase (d). The arrows denote the global movement of the flame surface, and the white line represents the calculated flame edge. Our primary focus was to calculate strain rate along the flame edge for different phases.

Figure 5.10 shows the strain rate PDFs along flame edge as the flame undergoes maximum pressure oscillation cycle, $p'_{max}(t)$, during thermoacoustic instability. For both equivalence ratios, the most probable strain rate is the lowest for phase (a) and highest for phase (b). At first the most probable strain rate on the flame front increased from phase (a) to phase (b) where it reached a maximum, then decreased back nearly to the value of phase (a) and again increased in the next cycle. Hence, at nodes of pressure perturbation cycle, the strain rate went down, and at antinodes strain rate went up. This occurred irrespective of the growth or decay of the combustion instability. However, the difference in strain rates between phases (b) and (a) or (d) and (c) is higher at near maximum pressure perturbation. This behavior of strain rate PDFs is similar for all equivalence ratios when $\phi < 0.5$, although the distributions are wider for the higher equivalence ratios due to higher frequency values of the first harmonic as seen in Fig. 5.4, particularly for those phases like (b) and (d) where maximum strain rates were observed.

5.4.8 Modeling Results

1D and 3D LEE were solved to analyze the self-excited combustion instability in the dual-chamber system shown in Fig. 5.1. Both ends of the combustor were closed, and the connecting nozzle was acoustically choked. Mean flow effects and entropy

Table 5.2 Comparison of instability modes from experiments and modeling

ϕ	Frequency (Hz)			Growth rate, rad/s
	Experimental	1D LEE (house code)	3D lee (COMSOL)	1D and 3D LEE
1.0	2170	2205	2192	0.005
0.5	1845, 2250	1903, 2297	1859, 2276	0.13
0.4	1740, 2250	1794, 2278	1755, 2267	0.29
0.3	1570, 2250, 4540	1612, 2281, 4587	1590, 2271, 4550	0.89
0.22	1430, 2250	1488, 2279	1452, 2270	1.9

Table 5.3 Identification of different instability frequencies associated with combustion instability of ultra-lean H_2/air in the dual-chamber system

Frequency (Hz)	Longitudinal	Transverse	Mixed
2170	1L		
1845	1L		
1740	1L		
2250		1T	
1570	1L		
4540			1L-1T
1430	1L		

waves were included in the calculations. The Mach number in the pre-chamber and the main chamber were negligible compared to that in the connecting nozzle. Crocco's pressure lag and Dowling's velocity lag models were implemented to model combustion response function (CRF). The pre-chamber and the nozzle were assumed filled with combustion products at adiabatic flame temperature. Additionally, the flame was assumed to sit in the main combustor. All thermochemical properties were calculated using the equilibrium module of COSILAB [53]. Since the connecting nozzle was acoustically choked, the pre-chamber did not influence the main chamber combustion stability. A sharp discontinuity was observed in all mode shapes at the flame location. In 1D LEE, pressure and temperature mode shapes were in phase with a minimal time lag.

Table 5.2 lists the calculated instability frequencies, and growth rate averaged over a single pressure oscillation cycle for the dual-chamber combustor. The results in Table 5.2 were obtained by using an in-house solver for 1D LEE and COMSOL 3D LEE. The results from 1D LEE overestimated the frequencies by 7–15%. However, the 3D model predicted the frequencies within 2–8% of the experimental values. For example, for $\phi = 0.3$, the first longitudinal and transverse modes (1L and 1T) occur at 1570 Hz and 2250 Hz, respectively. The mixed modes occur at frequencies higher than 4000 Hz. The following section identifies the instability modes associated with these frequencies.

Table 5.3 lists different frequencies associated with combustion instability in the system. The first frequency in FFT (Fig. 5.4) is always of the longitudinal mode of the combustor.

The ratio of the 1L mode to the other modes in FFT is always greater than 2. This indicates that the longitudinal mode is the dominant/strongest mode in the combustor. For any combustor, the longitudinal mode can easily be predicted using classical

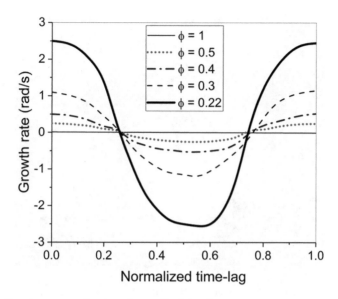

Fig. 5.11 Growth rates as a function of normalized time lag, τ

acoustic resonator theory (Eqs. 5.42 and 5.43). The higher-order modes are complex in nature. A careful investigation revealed that the transverse modes become dominant for lower equivalence ratios. The probable reason is that due to the shorter width of the combustor, transverse waves did not get enough time to interact with the flame for $\phi > 0.4$. However, at lower equivalence ratios, as the flame got weaker, transverse modes became important. The most interesting mode is the mixed mode, 1L-1T, when the longitudinal and transverse modes got coupled at a high frequency, 4540 Hz.

Figure 5.11 shows the growth rates as a function of normalized time lag, τ, for various equivalence ratios. The positive growth rate was observed when the time lag between pressure oscillations and heat release was in phase $\tau < 0.25$ or $\tau > 0.75$. For $\phi = 1.0$ the average growth rate was negligible, 0.005 rad/s. Compared to growth rates for $\phi \le 0.5$, damping was observed for $\phi = 1.0$ indicating that heat losses from the system were faster than the fluctuation to grow. This made the system stable for $\phi > 0.5$. However, below $\phi < 0.5$ flame dynamics got coupled with pressure oscillations and heat release, and growth rates increased. The growth rates reached maximum for $\phi = 0.22$.

5.4.9   Dynamic Mode Decomposition (DMD)

Dynamic mode decomposition (DMD) provides a better understanding of the dynamic behavior of the unsteady heat release, pressure perturbation, OH* signal,

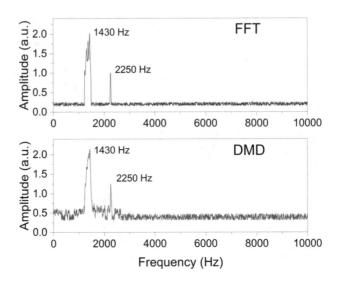

Fig. 5.12 Comparison of FFT pressure spectra with DMD spectra of OH* images

unsteady velocity, and vorticity field inside the combustor. The DMD algorithm resembles with proper orthogonal decomposition (POD), but DMD has several advantages over the POD. When POD contains the flow dynamics associated with multiple frequencies, every DMD model comprises of only a solo frequency. Thus, the physical significance of a particular frequency present in the flow field can be better understood using DMD rather than POD.

Dynamic mode decomposition (DMD) [28, 54] was employed to identify the dynamic behavior that has periodic occurrence in the dual-chamber combustor and to understand relations between the different physics. DMD mode spectrum was calculated using 500–1000 OH* images with a time separation of 45.5 µs. Figure 5.12 compares the DMD spectra with the pressure spectra calculated using FFT for $\phi = 0.22$. The 1L ($f = 1430$ Hz) and 1T modes ($f = 2250$ Hz) match well. Even the ratio of the amplitudes of 1L to 1T modes from FFT and DMD spectra agrees closely.

The spatial mode shapes of the DMD modes for $\phi = 0.22$ are presented in Fig. 5.13 for frequencies of 1430 Hz (1L) and 2250 Hz (1T). The DMD analysis was performed using mean subtracted images, so only the fluctuations in the OH* signal are visible. Figure 5.13a clearly shows the heat release region concentrated on the flame surface for $f = 1430$ Hz. This observation holds true for all other equivalence ratios. Thus, the first longitudinal (1L) mode always signifies the heat release mode of the combustor. Conversely, the first transverse (1T) mode as shown in Fig. 5.13b, $f = 2250$ Hz, is related to the transverse natural frequency of the combustor. While the 1L mode changes with the change in equivalence ratio (due to change in adiabatic flame temperature), 1T remains identical since it is a function of chamber geometry only.

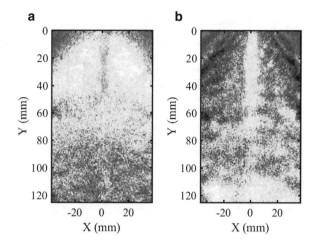

Fig. 5.13 DMD spatial mode at (**a**) $f = 1430$ Hz and (**b**) f = 2250 Hz from OH* images

5.4.10   Mechanisms Associated with Combustion Instability

Time-resolved simultaneous schlieren and OH* chemiluminescence imaging and modeling results from 3D LEE were utilized to investigate the mechanism behind thermoacoustic instability of ultra-lean premixed H_2/air combustion by pre-chamber hot jet ignition. The unstable modes of an absolutely unstable hot turbulent jet (jet temperature ranges between 900 K $< T <$ 1200 K [35]) acted as the initial sources for disturbance/perturbation during the initiation of ignition in the main chamber. Figure 5.14 schematically shows various sources of disturbances in the dual-chamber jet ignition combustion system. When the main chamber equivalence ratio went below $\phi < 0.5$, thermoacoustic instability triggered in. Flame speed and adiabatic flame temperature decreased significantly at ultra-lean conditions. Thus, the flame became progressively weak, and acoustic disturbances from the self-excited hot jet ignition and reflected pressure waves started affecting the flame dynamics. Strain rate along flame edge increased at antinodes of each pressure perturbation cycle and decreased at nodes. This oscillating strain rate (or in other words the fluctuating velocity field, u') supplied energy to form the coupling between unsteady heat release, q', and pressure fluctuations, p'. When losses through the system (heat losses, acoustic losses, damping, etc.) were higher than the growth rate (gain), instability started to decay. The overall energy in the coherent field depended upon the relative balance between these competing amplification and damping processes. The longitudinal modes were dominant for relatively higher equivalence ratio, $0.4 < \phi < 0.5$. But transverse and mixed modes become important for ultra-lean conditions, $\phi < 0.3$. Neverthless, the first longitudal mode remains the strongest mode of such thermoacoustic instability in a constant volume chamber.

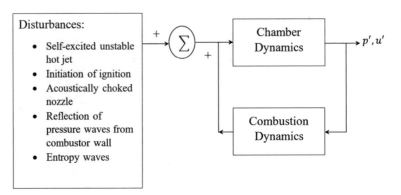

Fig. 5.14 Schematic of a dual-chamber jet ignition combustion system as a feedback amplifier

5.5   Conclusions

This chapter characterized thermoacoustic combustion instabilities of ultra-lean premixed H_2/air ignited by pre-chamber-generated hot turbulent jet using combined experiment and modeling. The key findings are summarized below:

1. The strongest mode always corresponds to the first longitudinal (1L) mode of the system. The first transverse (1T) mode is chamber geometry dependent. The higher-order modes are the complex coupled mode of the combustor. Unstable modes from 3D LEE match better with the experimental data as compared to 1D LEE. The first unstable mode in combustion instability always corresponds to the longitudinal mode of the system. Transverse and complex mixed modes arise at lower equivalence ratios.
2. A supercritical bifurcation occurs, and instability triggers in for $\phi < 0.5$. The frequency of the first longitudinal (1L) mode decreases with decreasing equivalence ratio due to change in adiabatic flame temperature. The frequency of the 1L mode ranges from 1400 to 2200 Hz.
3. Strain rates fluctuate along the oscillating flame edge. This causes higher strain at antinodes and lower strain at nodes of pressure perturbation cycle. The maximum strain rate reaches four to five folds of the minimum strain rate in a typical pressure perturbation cycle at the lean flammability limit for H_2/air. At the lean flammability limit, the fluctuating pressure reaches 25% of the mean pressure.
4. Dynamic mode decomposition (DMD) analysis was performed on OH* chemiluminescence images to identify the physical significance of the dominant unstable modes in the combustor. The 1L mode always corresponds to the heat release mode of the combustor.

References

1. United States, E.P.A: Office of transportation and air quality. In: Non-conformance Penalties for Heavy-Duty Diesel Engines Subject to the 2010 NOx Emission Standard. U.S. Environmental Protection Agency, Office of Transportation and Air Quality, Washington, DC (2012)
2. Peckham, J.: U.S. EPA doubles 'Non-conformance penalties' for Navistar 2012 diesel trucks. Dies Fuel News. **16**(34), 4–5 (2012)
3. Rapp, V., Killingsworth, N., Therkelsen, P., Evans, R., et al.: Chapter 4 Lean-Burn Internal Combustion Engines, pp. 111–146. Elsevier Inc. (2016) https://www.sciencedirect.com/science/book/9780128045572
4. Dunn-Rankin, D.: Lean Combustion: Technology and Control. Academic Press, Amsterdam/Boston (2008.) xi, 261 p., 8 p. of plates
5. Toulson, E., et al.: Visualization of propane and natural gas spark ignition and turbulent jet ignition combustion. SAE Int. J. Engines. **5**(4), 1821–1835 (2012)
6. Toulson, E., Watson, H., Attard, W.: Gas Assisted Jet Ignition of Ultra-Lean LPG in a Spark Ignition Engine, SAE Technical Paper 2009-01-0506. SAE (2009) https://doi.org/10.4271/2009-01-0506
7. Sadanandan, R., et al.: Detailed investigation of ignition by hot gas jets. Proc. Combust. Inst. **31** (1), 719–726 (2007)
8. Ghoniem, A.F., Oppenheim, A.K., Chen, D.Y.: Experimental and Theoretical Study of Combustion Jet Ignition. California University, Berkeley (1983)
9. Oppenheim, A., et al.: Jet Ignition of an Ultra-Lean Mixture, SAE Technical Paper 780637. SAE (1978) https://doi.org/10.4271/780637
10. Wolfhard, H.G.: The ignition of combustible mixtures by hot gases. J. Jet Propul. **28**(12), 798–804 (1958)
11. Lee, K., et al.: CO2 radiation heat loss effects on NOx emissions and combustion instabilities in lean premixed flames. Fuel. **106**, 682–689 (2013)
12. Quader, A.: Lean Combustion and the Misfire Limit in Spark Ignition Engines, SAE Technical Paper 741055. SAE (1974) https://doi.org/10.4271/741055
13. Clayton, R.M., Sotter, J.G., Woodward, J.W.: Injector response to strong, high-frequency pressure oscillations. J. Spacecr. Rocket. **6**(4), 504–506 (1969)
14. Natanzon, M.S., Culick, F.E.C.: Combustion instability. In: Progress in Astronautics and Aeronautics. American Institute of Aeronautics & Astronautics, Reston (2008)
15. O'Connor, J., Acharya, V., Lieuwen, T.: Transverse combustion instabilities: acoustic, fluid mechanic, and flame processes. Prog. Energy Combust. Sci. **49**, 1–39 (2015)
16. Fischbach, S.R.: Solid rocket motor combustion instability modeling in COMSOL multiphysics. In: COMSOL Conference. Boston (2015)
17. Prevost, M., Godon, J., Innegraeve, O.: Thrust oscillations in reduced scale solid rocket motors part I : experimental investigations. In: 41st AIAA/ASME/SAE/ASEE Joint Propulsion Conference & Exhibit Tucson, Arizona (2005) https://arc.aiaa.org/doi/abs/10.2514/6.2005-4003
18. Williams, F.A., Barrère, M., Huang, N.C.: Fundamental aspects of solid propellant rockets. In: AGARDograph, vol. iii-xxii, p. 791. Aerospace Research and Development of N.A.T.O. by Technivision Services, Slough/London (1969)
19. Price, E.W.: Review of experimental research on combustion instability of solid propellants, in solid propellant rocket research. Progr. Astronaut. Rocket. **1**, 561–602 (1960)
20. Price, E.W.: Combustion instability in solid propellant rocket motors. Astronaut. Acta. **5**(C), 63–72 (1959)
21. Bennewitz, J.W., Lineberry, D.M., Frederick, R.: Application of high-frequency pressure disturbances as a control mechanism for liquid rocket engine combustion instabilities. In: 9th AIAA/ASME/SAE/ASEE Joint Propulsion Conference. San Jose (2013)
22. Bennewitz, J.W., Lineberry, D.M., Frederick, R.A.: Investigation of a single injector with applied high frequency pressure disturbances for applications to liquid rocket engine

combustion instabilities. In: 49th AIAA/ASME/SAE/ASEE Joint Propulsion Conference, San Jose (2013) https://arc.aiaa.org/doi/pdf/10.2514/6.2013-3853
23. Zhang, H., et al.: Analysis of combustion instability via constant volume combustion in a LOX/ RP-1 bipropellant liquid rocket engine. SCIENCE CHINA Technol. Sci. **55**(4), 1066–1077 (2012)
24. Quinlan, M., et al.: Theoretical, Numerical, and Experimental Investigations of the Fundamental Processes that Drive Combustion Instabilities in Liquid Rocket Engines. Georgia Institution of Technology, Atlanta (2012)
25. Chehroudi, B.: Physical hypothesis for the combustion instability in cryogenic liquid rocket engines. J. Propuls. Power. **26**(6), 1153–1160 (2010)
26. Dranovsky, M.L., Vigor, Y., Culick, F.E.C.: Combustion Instabilities in Liquid Rocket Engines : Testing and Development Practices in Russia. American Institute of Aeronautics and Astronautics, Washington, DC (2007)
27. Temme, J.E., Allison, P.M., Driscoll, J.F.: Combustion instability of a lean premixed prevaporized gas turbine combustor studied using phase-averaged PIV. Combust. Flame. **161** (4), 958–970 (2014)
28. Lieuwen, T.C., Yang, V.: Combustion instabilities in gas turbine engines: operational experience, fundamental mechanisms and modeling. In: Progress in Astronautics and Aeronautics, vol. xiv, p. 657. American Institute of Aeronautics and Astronautics, Reston (2005)
29. Kim, D., et al.: Effect of flame structure on the flame transfer function in a premixed gas turbine combustor. J. Eng. Gas Turbines Power. **132**(2), 021502 (2010)
30. Steinberg, A.M., et al.: Flow–flame interactions causing acoustically coupled heat release fluctuations in a thermo-acoustically unstable gas turbine model combustor. Combust. Flame. **157**(12), 2250–2266 (2010)
31. Dowling, A.P., Stow, S.R.: Acoustic analysis of gas turbine combustors. J. Propuls. Power. **19** (5), 751–764 (2003)
32. Brookes, S.J., et al.: Computational modeling of self-excited combustion instabilities. J. Eng. Gas Turbines Power. **123**(2), 322 (2001)
33. Keller, J.J.: Thermoacoustic oscillations in combustion chambers of gas turbines. AIAA J. **33** (12), 2280–2287 (1995)
34. Biswas, S., et al.: On ignition mechanisms of premixed CH4/air and H2/air using a hot turbulent jet generated by pre-chamber combustion. Appl. Therm. Eng. **106**, 925–937 (2016)
35. Biswas, S., Qiao, L.: Prechamber hot jet ignition of ultra-lean H2/air mixtures: effect of supersonic jets and combustion instability. SAE Int. J. Engines. **9**(3), 1584–1592 (2016)
36. Möser, M.: Engineering Acoustics: an Introduction to Noise Control, vol. xi, p. 289. Springer, Berlin/New York (2004)
37. Fahy, F.: Foundations of Engineering Acoustics, vol. xix, p. 443. Academic, San Diego (2001)
38. COMSOL.: COMSOL multiphysics 4.0 (2015)
39. Sagaut, P.: Large Eddy Simulation for Incompressible Flows an Introduction, 3rd edn. Springer Books, New York (2006)
40. Frezzotti, M.L., et al.: Response Function Modeling in the Study of Longitudinal Combustion Instability by a Quasi-1D Eulerian Solver (2015)
41. Monkewitz, P.A., et al.: Self-excited oscillations and mixing in a heated round jet. J. Fluid Mech. **213**(1), 611 (2006)
42. Yu, M.H., Monkewitz, P.A.: The effect of nonuniform density on the absolute instability of two-dimensional inertial jets and wakes. Phys. Fluids A Fluid Dyn. **2**(7), 1175–1181 (1990)
43. Monkewitz, P.A., Sohn, K.: Absolute instability in hot jets. AIAA J. **26**(8), 911–916 (1988)
44. Raynal, L., et al.: The oscillatory instability of plane variable-density jets. Phys. Fluids. **8**(4), 993–1006 (1996)
45. Kinsler, L.E.: Fundamentals of Acoustics, 4th edn. Wiley, New York (2000)
46. Rayleigh, J.W.S., Lindsay, R.B.: The Theory of Sound, 2nd edn. Dover Publications, New York (1945)
47. Reaction Design.: Reaction Workbench 15131 San Diego (2013)
48. ANSYS.: ANSYS Fluent Academic Research, Release 15.0 (2015)

49. Guo, H., Chen, Y.: Supercritical and subcritical Hopf bifurcation and limit cycle oscillations of an airfoil with cubic nonlinearity in supersonic\hypersonic flow. Nonlinear Dyn. **67**(4), 2637–2649 (2012)
50. Biswas, S., Qiao, L.: A comprehensive statistical investigation of Schlieren Image Velocimetry (SIV) using high-velocity helium jet. Exp. Fluids. **58**(3), 1–20 (2017)
51. Kundu, P.K., Cohen, I.M., Dowling, D.R.: Fluid Mechanics, vol. xxvi, 5th edn, p. 891. Academic Press, Waltham (2012)
52. Westerweel, J.: Digital Particle Image Velocimetry – Theory and Application. Delft University, Delft (1993)
53. COSILAB.: Inc (2017)
54. Speziale, C.G., Sarkar, S., Gatski, T.B.: Modeling the pressure-strain correlation of turbulence - an invariant dynamical systems approach. J. Fluid Mech. **227**, 245–272 (1991)

Chapter 6
Ignition by Multiple Jets

Contents

6.1 Background of Multi-jet Ignition

The main reason that hot turbulent jet ignition has become attractive to gas engine manufacturers is that hot jet ignition can achieve faster burn rates due to the ignition system producing multiple, distributed ignition sites, which has greater likelihood igniting a lean mixture compared to spark ignition. This leads to better thermal efficiency and low NOx production. Compared to conventional spark ignition, a hot jet has a much larger surface area leading to multiple ignition sites on its surface which can enhance the probability of successful ignition and cause faster flame propagation and heat release. Over the last few decades, pre-chamber jet ignition had technologically advanced from conceptual design phase to actual engines. The early designs developed by Gussak [1–4], Oppenheim [5, 6], Wolfhard [7], and Murase [8] showed the promise of lean ignition by a hot turbulent jet. Later, Ghoneim and Chen [9], Pitt [10], Yamaguchi [11], Elhsnawi [12], Sadanandan [13], Toulson [14, 15], Gholamisheeri [16], Attard [17], Perera [18], Carpio [19], Shah [20], Karimi [21], Thelen [22], and Biswas [23, 24] further investigated in detail the parametric effects and fundamental physics of turbulent jet ignition in laboratory scale prototype combustors and at engine-relevant conditions. All these studies support that turbulent jet ignition possesses several advantages over traditional

© Springer International Publishing AG, part of Springer Nature 2018   129
S. Biswas, *Physics of Turbulent Jet Ignition*, Springer Theses,
https://doi.org/10.1007/978-3-319-76243-2_6

spark ignition during ultra-lean combustion such as higher ignition probability, faster burn rates, and multiple ignition kernels.

Since the hot turbulent jet plays a key role in turbulent jet ignition (TJI), the geometric and thermodynamic parameters associated with the jet behavior have been studied by several researchers. Attard et al. [25] visualized the in-cylinder combustion process at different levels of dilution initiated by six turbulent hot jets. They found TJI produced rapid combustion compared to conventional spark ignition. Zhang et al. [26] investigated multi-jet ignition for diesel premixed compression ignition system and found that the spark timing is an effective way to optimize the emissions and thermal efficiency. Gentz et al. [27] studied the influence of orifice diameter on TJI system employing various geometry and configuration nozzle plates with a number of nozzles/orifices varying from 2 to 6. They found at near stoichiometric conditions the burn durations were shortest for the multiple nozzles compared to single nozzles.

Alongside the fundamental research conducted in universities and national laboratories, engine manufacturers have invested efforts to implement TJI into real engines. Over the last several decades, engine companies like Bosch [28], Ford Motors [29], GM [30], Dresser-Rand [31], Rolls-Royce [32], Mahle Powertrain [33], Woodward [34], and Caterpillar [35] have come up with different pre-chamber designs for hot turbulent jet ignition. Some selected pre-chamber designs are shown in Fig. 6.1. Robert Bosch Gmbh (1980) patented a design named external ignition combustion device for IC engines, which was essentially a spark-ignited pre-chamber design. The ignition in the main engine combustor relied on the vortices that would be generated after the jet hits the piston head crevice and ignites the chamber [28]. Ford Motor Company (1977) designed a pre-chamber capable of producing a swirling flame jet to provide superior mixing with the unburned combustible mixture in the main combustor, particularly useful for a rotary engine [29]. General Motors Corporation (1995) designed a torch jet spark plug to enhance the burning rate within the main chamber [30]. Dresser-Rand Company (1996) conceptualized a walled precombustion chamber unit to mount on the cylinder head sidewall. This design also contained multiple nozzles connecting the pre-chamber to the squish volume [31]. Rolls-Royce Marine (2011) designed a pre-chamber unit for an SI engine that had four holes around the boundary, directed 45 downward [32]. This pre-chamber was arranged to be placed in the upper part of the engine's cylinder so the gas jets could come at an angle of 45° to the piston head. Mahle Powertrain, LLC (2012) designed an ignition system for an internal combustion engine where the ignition system included a housing, an ignition device, an injector, and a pre-chamber with six nozzles equally spaced around the pre-chamber tip [33]. Woodward, Inc. (2014) proposed a multi-chamber, multi-nozzle igniter for rotary and propeller type aircraft applications [34]. They used multiple orifices with varying diameters in their pre-chamber. Caterpillar Inc. (2014) also designed the pre-chamber combustion tip with 5–13 holes [35]. Oppenheim [36] proposed a pre-chamber with multiple nozzles mounted at the cylinder head and recommended that the number of jets anywhere between 1 and 5 would serve best for main chamber combustion. Most of the abovementioned TJI systems were designed for gasoline and natural gas.

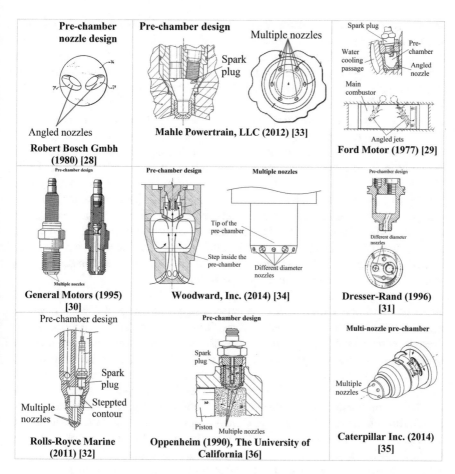

Fig. 6.1 Existing pre-chamber designs with multiple nozzles

There are several commonalities between all these existing pre-chamber designs shown in Fig. 6.1. Nearly all used multiple jets for ignition. Additionally, the nozzles are oriented in the pre-chamber in such manner so that the emerging hot jets are directed toward the potential hot pockets in the main combustion chamber where ignition is more probable. In the squish volume (clearance volume) of an engine, the normal distance between the piston head to the cylinder head is usually short. Thus, to negate any boundary/wall effect, the jets are released at an angle of 40–70° normal to the piston motion. This allows more time for the jets to interact with the charge so that the ignition can occur easily. Secondly, for all the designs presented in Fig. 6.1, the pre-chamber is a cylindrical volume with a spark on the top or side, except that Woodward and Rolls-Royce used a pre-chamber that had multiple area changes to increase vorticity and turbulence inside the pre-chamber. Nevertheless, for all these pre-chambers, the length to diameter ratio, L/D, ranged from 2 to 6. In fact, this pre-chamber design parameter has a deterministic effect on the pattern of the hot jets issued from the pre-chamber through multiple nozzles, which subsequently determines the ignition behavior in the main chamber by the multiple hot jets. This will be discussed in detail later.

However, not only commonalities, there exist dissimilarities in the pre-chamber designs as well depending on the engine manufacturer. For example, the pre-chamber location varies a lot depending on the application and engine type. While Mahle and Bosch placed the pre-chamber at the center of the cylinder head, Ford and GM had the pre-chamber at a corner location near the exhaust valve to assure that the charge around the exhaust valve had already been consumed when the exhaust valve would open. Another design aspect is the diameter of the connecting nozzles. While most of the designs opted for a constant diameter for all the nozzles, Dresser-Rand used different diameters based on nozzle location in the pre-chamber. The idea of using different nozzle diameters at the different pre-chamber location was based on from the observed engine performance testing.

Even though the fundamental concept behind these designs is the same, the varied pre-chamber designs would certainly have a major effect on the main chamber ignition characteristics. As mentioned earlier, one thing that is common in all these pre-chamber designs is that the number of pre-chamber nozzles/orifices is always more than one. In practical engines, the number of nozzles was estimated in an ad hoc way, based upon comparing the engine performance with a different number of pre-chamber nozzles. Previous studies demonstrated that increasing the number of turbulent hot jets results in faster flame propagation and thus higher combustion efficiency. Nevertheless, there is a lack of fundamental understanding of the effect of multiple jets on the ignition and combustion behavior of ultra-lean mixtures. A better understanding of multi-jet ignition is required, e.g., jet interactions, ignition pattern/mechanism, the effect of pre-chamber spark location on jet characteristics, and so on. This motivated us to explore the ignition phenomena by multiple jets generated by pre-chamber combustion. We wanted to understand the effectiveness of multi-jets compared to a single jet, how jets interact with each other, and the influence on main chamber ignition characteristics such as the burn rate and ignition delay. One of the key questions is whether each individual jet in a multi-jet system contributes to the ignition process in the same way. This fundamental understanding would be helpful for future pre-chamber designs in terms of optimization of the pre-chamber and the number of jets to achieve the best performance.

In all previous studies, the number of pre-chamber nozzles, nozzle geometry, and configuration was estimated in an ad hoc way, based upon comparing the combustion performance (such as ignition delay, burn time, flame surface area, turbulent flame speed, etc.) with different nozzle configurations. The current study explores the effectiveness of multiple jets compared to a single jet and the role of an individual jet in a multi-jet system. The current study focuses on answering the following fundamental questions, e.g., what is the role of an individual jet in a multi-jet system? Do all the individual jets behave/contribute in the same way? How does the jet interaction affect the ignition dynamics? How ignition characteristics of multi-TJI such as ignition probability and ignition delay compare with the single TJI? What are the geometric or thermophysical factors that govern the ignition mechanism by multiple jets? Additionally, the present work provides important insight into pre-chamber design and optimization. Pre-chamber equivalence ratio and spark location combined to play a critical role in determining the effectiveness of different multi-nozzle configurations. Since the success of TJI by multiple jets lies in the optimized pre-chamber design, these results provide useful information for future pre-chamber design to achieve high efficiency.

This chapter is focused on the ignition characteristics of ultra-lean H_2/air mixtures by multiple hot turbulent jets. Two different multi-jet orientations were used, straight and angled. For each orientation, three inline jets, one center jet and two other jets, one on each side of the center jet, were used. For the straight orientation, all three jets were parallel to each other. For the angled orientation, the center jet was straight while the side jets were angled. We also studied the effect of spark location inside the pre-chamber and pre-chamber and main chamber fuel/air equivalence ratio on ignition characteristics such as the flammability limit and ignition probability. In this chapter, we first compare the multiple jet ignition with the single jet ignition. Then we compare straight and angled multi-jets and assess how they affect the ignition delay and ignition probability of the main combustor mixture. Lastly, we numerically simulated the flame propagation process in the pre-chamber to gain further insight that helps to explain the experimental observations.

6.2 Experimental Method

The experimental setup has been discussed in detail in Chap. 2. A small volume, 100 cc cylindrical stainless steel (SS316) pre-chamber was attached to the rectangular (43.18 cm × 15.24 cm × 15.24 cm) carbon steel (C-1144) main chamber. The pre-chamber was a cylindrical volume 8.9 cm (3.5 inches) long and 3.81 cm (1.5 inches) in diameter. The main chamber to pre-chamber volume ratio was 100:1. Length scales of the pre-chamber and the main chamber used in the present study were chosen to ensure the pre-chamber to main chamber volume ratio is consistent with those in large bore stationary natural gas engines. We used the same dual combustion chamber set up with different nozzle plates. However, the present study differed from the previous one in terms of the number of hot turbulent jets igniting the main combustion chamber. In the current study, instead of one single jet, multiple hot turbulent jets were generated from pre-chamber combustion. Figure 6.2a shows the schematic of the dual combustion chamber with multiple (three) jets. Figure 6.2b shows two different multi-nozzle plates – straight nozzle plate and angled nozzle plate – used in the experiment. Each multi-nozzle plate contained three-nozzle holes: one center nozzle and two other nozzles, one on each side of the center nozzle. The spacing between each pair of nozzles was 12.7 mm (0.5″) as shown in Fig. 6.2b. The nozzle diameter and nozzle length were 2.54 mm

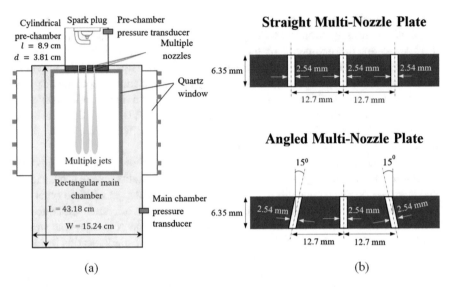

Fig. 6.2 (a) Schematic of the dual combustion chamber with multiple hot turbulent jets, (b) two different multiple nozzle plates, straight (top) and angled (bottom)

(0.1″) and 6.35 mm (0.25″), respectively, and were identical for all three nozzles. For the angled multi-nozzle plate, the center nozzle was straight. The side nozzles were angled outward with an angle of 15° to the vertical line. For the angled multi-nozzle plate, the length of the side nozzles was slightly higher, $l_{side} = 0.25″/\cos(15°)$. However, the jet exit locations in the main chamber were kept identical for both multi-nozzle plates.

The pre-chamber mixture was maintained at equivalence ratio, $\phi = 1.0$ for most of the tests. However, to study the effect of pre-chamber equivalence ratio on the main chamber ignition behavior, we explored off-stoichiometric pre-chamber conditions as well. We varied the pre-chamber equivalence ratio from lean, $\phi = 0.8$, to rich, $\phi = 1.3$. As will be shown later, the hot jet characteristics greatly depend on the pre-chamber equivalence ratio. The equivalence ratio in the main chamber was varied from $\phi = 0.5$ to the lean flammability limit of H_2/air around $\phi = 0.3$. Industrial grade H_2 (99.98% pure) was used for all tests. From two different pre-mixers, premixed H_2/air of dissimilar equivalence ratios were fed to pre-chamber and the main chamber separately. To separate the pre-chamber and the main chamber with different equivalence ratios, a lightweight, 25 ± 1.25 μm thick aluminum diaphragm was used. The ruptured diaphragm was replaced after each test. Details of the diaphragm rupture process have been discussed in our previous work [23].

A spark created by a 0–40 kV capacitor discharge ignition (CDI) system ignited the pre-chamber H_2/air mixture. Four different spark locations were tested in our experiment as shown in Fig. 6.3a. The spark position was measured from the nozzle exit as shown by the reference $x - y$ coordinates in Fig. 6.3a. Figure 6.3b shows the

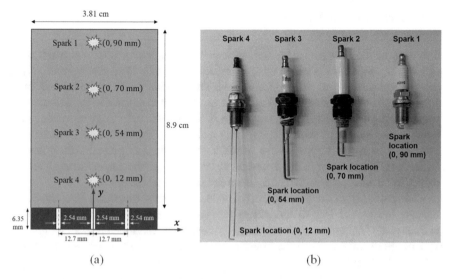

Fig. 6.3 (**a**) Schematic of the four different spark locations inside the pre-chamber, (**b**) four different spark plugs used in our experiment

actual spark plugs with customized electrodes. A spark gap of 1.5 mm was maintained for all the spark plugs. The measured spark energy was 120 mJ with a 10% uncertainty range. At the beginning of the experiment, both chambers were kept quiescent. In other words, our study did not consider main chamber turbulence induced by piston motion in a real engine. The purpose was to reduce the complexity of the system and focus on the physics of the hot jet issued from the pre-chamber. As soon as the spark ignited the pre-chamber mixture, a flame propagated inside the pre-chamber and entered the main chamber in the form of hot turbulent jets through multiple nozzles. Combustion in the main chamber took place at constant volume.

Simultaneous high-speed schlieren and OH* chemiluminescence imaging were utilized to visualize the evolution of the hot turbulent jets as well as the ignition process in the main chamber. A Z-type schlieren system consisted of a 100 Watt (ARC HAS-150 HP) mercury vapor UV light source with a condensing lens, two concave parabolic mirrors (15.24 cm or 6 inches in diameter, focal length 1.2 m), and a high-speed Phantom v7 camera to capture the ignition event with a resolution of 800 × 720 pixels with a frame rate up to 15,000 fps. The high-speed OH* chemiluminescence imaging was used to find the traceable amount of OH* radicals present in the hot jets as well as to identify the main chamber ignition location and flame propagation processes. A second high-speed Phantom v7 camera along with a gated image intensifier (VS4-1845HS) with 105 mm UV len was utilized to detect OH* signals at a very narrow band 386 ± 10 nm detection limit.

6.3   Results and Discussion

The findings from multi-jet ignition of lean H_2/air mixtures were compared with single jet ignition results which were described in Chap. 2. Here two separate single jets were chosen for comparison purposes. The first single jet was produced by a single nozzle with a diameter of 4.5 mm. It was chosen to ensure the total area of the single nozzle was the same as the total area of the three-nozzle multi-jet system with each nozzle having a diameter of 2.54 mm (or 0.1 inch). The second single nozzle we used for comparison purposes had a diameter of 2.6 mm, nearly equal to the diameter of the individual nozzle in the three-nozzle system. Both the single nozzles had the length to diameter, l/d, ratio of 2.5, same as the multi-nozzles.

The results and discussion section is divided into four subsections. First, the effect of spark location on lean flammability limit, ignition probability, and ignition delay is discussed. The performance and contribution of each individual jet in a multi-jet system are examined. Then the effect of the pre-chamber equivalence ratio on the ignition characteristics is studied. Afterward, the ignition characteristics of lean H_2/air mixtures by straight multi-jets are compared with angled multi-jets. Lastly, results from the numerical simulations of the pre-chamber flame are used to explain the experimental observations in terms of the performance/contribution of the individual jet in the multi-jet system.

6.3.1   Effect of Pre-chamber Spark Location

The pre-chamber spark location has a deterministic effect on the ignition characteristics of the main chamber ultra-lean H_2/air mixture by multiple jets. It was found that depending on the spark location inside the pre-chamber, the jet ignition characteristics in the main chamber can vary drastically. The effect of spark location on lean flammability limit, ignition probability, ignition delay, and relative importance of the individual jets in the multi-jet system is discussed in the following.

6.3.1.1   Lean Flammability Limit

In the present study, the lean flammability limit is the minimum fuel condition that has an ignition probability of 100%. Ignition probability can be expressed by Eq. (6.1), which is the ratio of the successful number of ignition events, $N_{ignition}$ to the total number of test runs, N_{total}.

$$\text{Ignition Probability} = \frac{N_{ignition}}{N_{total}} \tag{6.1}$$

Figure 6.4 shows the lean flammability limit of H_2/air by multiple jets using a straight multi-nozzle plate at various spark locations. The multi-jet lean limits are

Fig. 6.4 Flammability limit of H_2/air from multi-jet ignition compared to a single jet

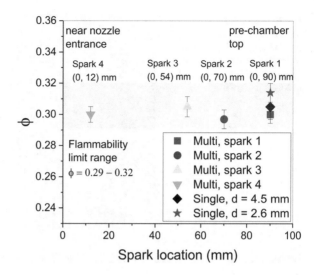

compared with those of the single jets. For all spark locations, the lean flammability limit of H_2/air ignited by straight multiple jets lies within a small range, $\phi_{\text{limit}} \approx 0.3 - 0.31$. The lean flammability limits for the single jets with diameters of 4.5 mm and 2.6 mm are $\phi_{\text{limit}|4.5\ \text{mm}} \approx 0.31$ and are $\phi_{\text{limit}|2.6\ \text{mm}} \approx 0.32$, respectively. Note that the spark was located on top of the pre-chamber for both single jets (spark 1). Thus, single or multiple jet systems show similar lean flammability limits, given that the total or individual nozzle area of the two systems is the same. In other words, multiple jets do not further reduce the lean flammability limit of H_2/air in the main chamber.

6.3.1.2   Ignition Probability

Even though the lean flammability limit remains similar for single and multiple hot jets, ignition probability of multiple jets changes near the lean flammability limit. Figure 6.5 displays the ignition probability at the lean limit for multiple jets versus single jet. As we go below the lean flammability limit, the mixture suddenly becomes unignitable. There exists a zone, $\phi = 0.26 - 0.3$, where ignition is possible, but with a probability less than 100%. Engine manufacturers do not wish to operate in this range since ignition probability is not 100%. It is evident from Fig. 6.5 that the ignition probability increased significantly for multiple jet ignition. The ignition probability was higher for multi-jets compared to a single jet. For example, for spark position 2, at $\phi = 0.285$, the ignition probability was 82%. At the same equivalence ratio, a single nozzle with a diameter of 2.6 mm and 4.5 mm had an ignition probability of 17% and 28%, respectively. The likely reason for higher ignition probability is because of cumulative ignition probability. The ignition mechanism for single jet and multi-jet is identical and governed by the nondimensional ignition

Fig. 6.5 Ignition
probability near the
flammability limit for multi-
jet ignition with 4 different
spark locations as compared
to single jet ignition

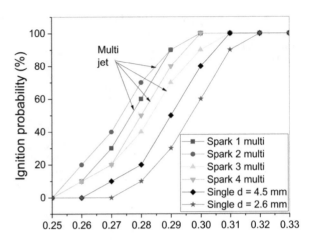

Damköhler number. Details of the ignition Damköhler number calculation is reported in our earlier work [23, 37]. However, as the number of jets increased, the cumulative probability of achieving the ignition Damköhler number increased as well. The pre-chamber spark location also affects the ignition probability. Spark locations 1 and 2 have higher ignition probability compared to the spark locations 3 and 4. Greater ignition probability for higher spark location is consistent with Thelen et al.'s observations [22]. Thelen et al. found that locating the spark high in the pre-chamber produces a jet that is more effective at quickly igniting the main chamber. For a spark location further away from the nozzle, most of the pre-chamber mixture gets burned and creates a higher-pressure differential to produce higher turbulence level and better mixing in the main chamber. This helps in faster flame propagation and higher ignition probability. A higher ignition probability from multiple jet ignition would allow an engine manufacturer to operate at the lean flammability limit with more confidence, even with a little variability or uncertainty associated with engine's fuel injection system [38].

6.3.1.3 Ignition Delay

Figure 6.6 shows the time histories of pressure recorded in the pre-chamber and the main chamber for two different spark locations, spark 1 and spark 3, respectively. The main chamber equivalence ratio was kept at $\phi = 0.4$, while the pre-chamber mixture was stoichiometric. The intention was to see if there were any differences in the main chamber combustion processes by changing the spark position in the pre-chamber. The pre-chamber pressure profile for two different spark locations, spark 1 and 3, matches quite well within experimental uncertainty. However, the ignition delay in the main chamber is different. The spark located near the nozzle entrance in the pre-chamber (spark 3) has a lower ignition delay compared to the spark that is farther away from the nozzle entrance (spark 1). The difference in ignition delay for $\phi = 0.4$ is approximately 1.96 milliseconds. This is because the

Fig. 6.6 Typical pressure profiles in pre-chamber and main chamber for two different spark locations inside the pre-chamber. The main chamber mixture was at $\phi = 0.4$

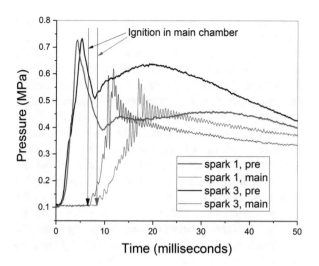

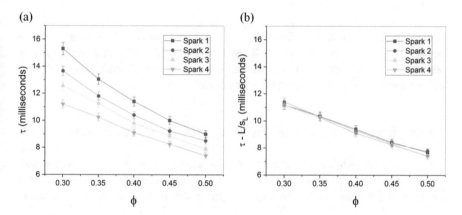

Fig. 6.7 (**a**) Ignition delay variation with equivalence ratio in the main chamber for different spark locations, (**b**) $(\tau - L/s_L)$ variation for different spark locations

pre-chamber flame had to travel less distance when the spark location was near the nozzle entrance. A detailed discussion on the ignition delay resulting from various spark locations will be presented in the following section. The readers may notice oscillations in the main chamber pressure profile around the peak pressure. This is due to thermoacoustic oscillations in the main chamber at lean operating conditions, which was discussed in our previous work [24].

Main chamber ignition delay as a function of equivalence ratio for four different spark positions is shown in Fig. 6.7a. As expected the ignition delay increases with decreasing equivalence ratio. For the spark position 1, farthest from nozzle entrance, ignition delay changes from 9 milliseconds at $\phi = 0.5$ to 15.3 milliseconds at $\phi = 0.3$. The other spark positions follow a similar trend. For a specific equivalence ratio, the ignition delay increased with the spark location farther away from the nozzle entrance. Thus, spark 1 had higher ignition delay compared to spark 2, 3, and 4. For example, at $\phi = 0.5$, spark 1 had an ignition delay of 9 milliseconds, spark 2 had an ignition delay of 8.5 milliseconds, spark 3 had 7.9 milliseconds, and spark 4 had 7.4 milliseconds. The reason is already explained in the previous section when discussing Fig. 6.4. The slight change in ignition delay was caused due to the fact that the pre-chamber flame had to travel a different length to enter into the main chamber. It is true that ignition did not occur as soon as the hot turbulent jet entered the main chamber. Rather, the jet accelerated initially and then decelerated, and ignition initiated from the deceleration phase of the jet.

To establish our claim that the difference in ignition delay was due to the different path lengths traveled by the pre-chamber flame, we plotted the ignition delay from a common reference position. We chose spark 4, the nearest position to the nozzle entrance as our reference position. Assuming laminar flame propagation within the pre-chamber, we calculated the time taken by the flame to travel to this reference position. Then, we subtracted this time from the respective ignition delays from spark positions 1, 2, and 3. Thus, we plotted the quantity, $(\tau - L/s_L)$, instead of the ignition delay, τ, as a function of equivalence ratio in Fig. 6.7b. Here L is the pre-chamber length in the direction of flame propagation and s_L is the laminar burning speed. All ignition delays collapsed onto one curve. This proved the ignition delays differed for various spark locations because the pre-chamber flame traveled different lengths depending on the spark position.

6.3.1.4   Behavior of Individual Jets in the Multi-jet Ignition System

The high-speed schlieren technique enabled visualization of the jet penetration, ignition, and the subsequent turbulent flame propagation processes in the main chamber. High-speed OH* chemiluminescence was used to identify the onset of ignition in the main chamber. Figure 6.8a–d show a time sequence of the simultaneous schlieren and OH* chemiluminescence images of the ignition processes in the main chamber for four different spark positions.

It is interesting to notice in Fig. 6.8a, b that for spark position 1 and 2, the main chamber ignition started from the side jets, as seen clearly from the OH* signals. After a while, the main chamber was first ignited by the two side jets, the center jet entered a little later and then further assisted in the main chamber ignition process. This is evident since the OH* signal intensity increased as the center jet participated in the ignition process. For spark position 1, the two side jets started the ignition at 11.4 milliseconds. However, the center jet enters the main chamber at around 13.4 milliseconds, and then the OH* signal intensity increased. As we changed the spark position closer to the nozzle entrance, such as spark position 3 as shown in Fig. 6.8c, ignition of the main chamber mixture was caused by all three jets nearly simultaneously. Among all three jets, the OH* intensity was higher for the center jet as compared to the side jets. Nevertheless, all three jets took part to initiate ignition in the main chamber.

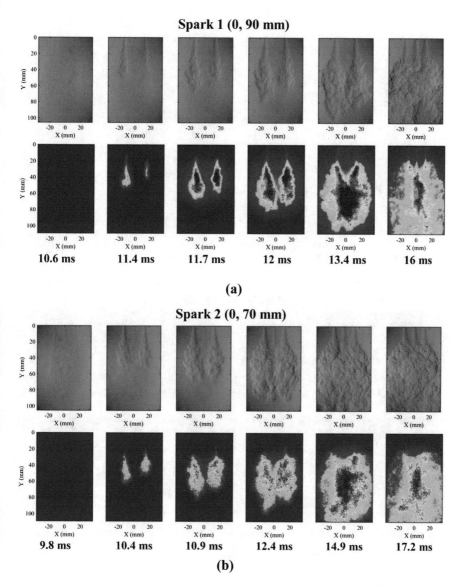

Fig. 6.8 The time sequence of simultaneous schlieren (top) and OH* chemiluminescence (bottom) images showing multi-jet ignition process in the main chamber for H_2/air at $\phi = 0.4$ (main chamber mixture) using the straight multi-nozzle plate with four different spark positions, (**a**) spark 1, (**b**) spark 2, (**c**) spark 3, (**d**) spark 4

Figure 6.8d shows an entirely different style of ignition. Note in this case; the spark location was very close to the nozzle entrance. First, a laminar flame jet entered through the center nozzle followed by two other laminar flame jets through the side nozzles. These laminar flame jets soon merged into one and progressed nearly at the

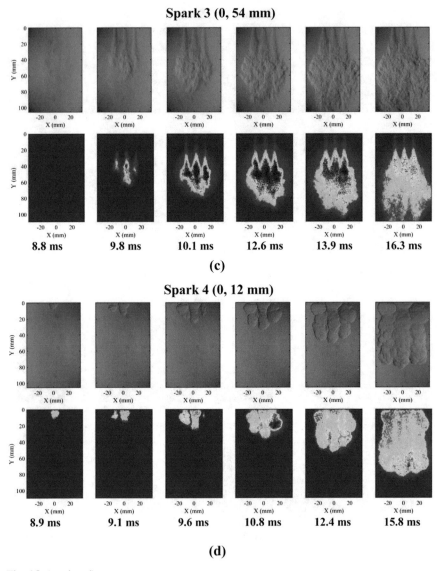

Fig. 6.8 (continued)

laminar flame speed of the main chamber H$_2$/air mixture. The longitudinal rate of flame propagation in the main chamber was much faster compared to the transverse flame propagation. While all other spark locations (spark 1–3), initiated main chamber ignition by turbulent jets, spark 4 alone started the ignition by a laminar or slightly turbulent flame jets. Note that spark 4 was only 12 mm away from the nozzle entrance. This phenomenon will be elaborated and explained later in the computational results section.

6.3.2   Effect of Pre-chamber Equivalence Ratio

6.3.2.1   Regime Diagram for Multi-jet Ignition

Figure 6.8 shows an interesting finding that depending on the pre-chamber spark location either the side jets, the middle jet, laminar or slightly turbulent flame jets, or all the jets would ignite the main chamber mixture. Also, the turbulence level of the hot jet depended on the spark position. To understand how other pre-chamber conditions than spark positions affect the main chamber ignition process, we examined the effect of pre-chamber equivalence ratio. Since the pre-chamber mixture should not be too off from the stoichiometric condition, we varied the pre-chamber equivalence ratio from $\phi = 0.8$ to 1.3.

Depending on the pre-chamber equivalence ratio and pre-chamber spark position, the side jet, or the middle jet, or all the jets can initiate ignition in the main chamber. To make sure the observations were repeatable, we did five experiments for each equivalence ratio for every spark location. Figure 6.9 shows the multi-jet ignition pattern produced by the straight multi-jet nozzle plate for four different spark locations and different pre-chamber equivalence ratios. We found the existence of four different zones based on which jet/jets initiated main chamber ignition. For all the pre-chamber equivalence ratios, for spark position 1, ignition happened through the side jets. For spark position 2, ignition occurred by the side jets for $0.8 < \phi < 1.15$. For $1.15 < \phi < 1.3$, ignition happened by all the jets. For spark position 3, for $0.8 < \phi < 0.98$, ignition happened through the middle jet. However, as we increased the fuel concentration, for $0.98 < \phi < 1.3$, ignition was initiated by all the jets. For spark position 4, ignition initiated through laminar or slightly turbulent flame jets.

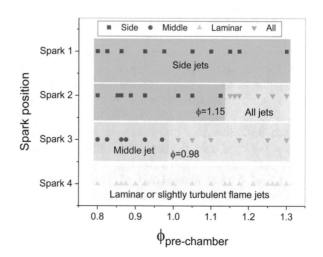

Fig. 6.9 A regime diagram for multi-jet ignition using straight multi-nozzle plate based on which jet initiated the main chamber ignition for different spark positions and pre-chamber equivalence ratio

6.3.2.2 Reynolds Number

To characterize the turbulence due to the hot jet, the Reynolds number of the jet was calculated at the nozzle exit using the following relation:

$$Re = \frac{\rho U_j d}{\mu} \qquad (6.2)$$

where U_j is the exit jet velocity at the nozzle exit, d is the nozzle diameter, ρ and μ are, respectively, the density and viscosity of the pre-chamber combustion products calculated by numerical simulations discussed in the subsequent section. Reynolds number varied with the nozzle diameter, the pressure differential between the two chambers, Δp and pre-chamber equivalence ratio. Nozzle diameter was constant throughout. However, the pre-chamber equivalence ratio and Δp were varied. The pressure differential Δp was a function of spark position. Spark located farther away from the nozzle entrance consumed more pre-chamber mixtures and thus had a higher Δp. Since the discharging jet was driven by the pressure differential between the pre-chamber and the main chamber, the nozzle exit velocity was higher for the farthest spark position. Thus, the nozzle exit velocity was higher for spark 1 compared to spark 2 and so on. For stoichiometric pre-chamber, for spark position 1 the Reynolds number was 57,000, and for spark position 4 the Reynolds number was 3400. Thus, the spark location played a key role determining the turbulence level of the discharging jet. For different pre-chamber stoichiometric conditions, for spark position 1, the Reynolds number was in the range of 43,000–67,000. The Reynolds number decreased as the spark position became closer to the nozzle. For spark position 2 and 3, the range of Reynolds number was 12,000–37,000. For spark position 4, the range of Reynolds number lied between 2500 and 8000. Thus, for spark position 4, ignition initiated through laminar or slightly turbulent flame jets.

6.3.2.3 Switching Ignition Pattern

The relative importance of the individual jet in a multi-jet system changes with pre-chamber equivalence ratio given a fixed spark position. For spark location 3, ignition pattern switched for a pre-chamber equivalence ratio of $\phi_{pre} = 0.98$. A center jet initiated the ignition in the main chamber for $\phi_{pre} < 0.98$, while for $\phi_{pre} > 0.98$ all the jets simultaneously initiated ignition in the main chamber. Figure 6.10 illustrates this switching ignition pattern as a function of ϕ. For $\phi_{pre} = 0.82$ and 0.96, the center jet initiated the main chamber ignition as evident from OH* images. However, for higher pre-chamber equivalence ratios, $\phi_{pre} = 1.1$ and 1.3, all the jets initiated the main chamber ignition.

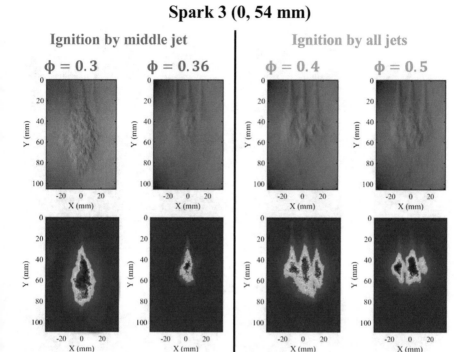

Fig. 6.10 The transition of multi-jet ignition pattern by changing pre-chamber equivalence ratio at spark location 3

6.3.3 Effect of Angled Jets Versus Straight Jets

In the following section, the ignition characteristics of lean H_2/air by angled multiple hot turbulent jets are discussed. Again, high-speed schlieren and OH* chemiluminescence were used for visualization in the main chamber.

6.3.3.1 Ignition Characteristics by Angled Multi-jets

Figure 6.11a–d illustrate the ignition time sequence in the main combustion chamber by angled multi-jets. In the angled multi-jet system, the side jets were angled 15° from the center jet. Similar conclusions for straight multi-jets were found for angled multi-jet ignition in terms of which jet was initiating the ignition in the main chamber. In angled multi-jet ignition, for spark position 1 and 2, side jets initiated the main chamber ignition. For spark position 3, all the jets initiated the ignition nearly simultaneously. Ignition occurred through propagating laminar or slightly turbulent flame jets for spark position 4. However, since the side jets were slightly angled, individual flame fronts took a longer time to merge with each other. Except that, the ignition pattern was identical for straight and angled multi-jets.

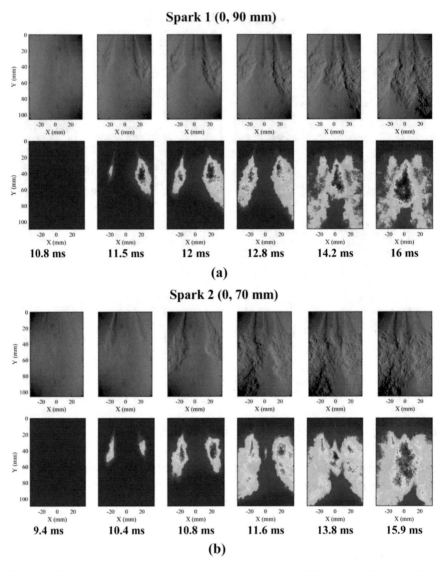

Fig. 6.11 The time sequence of simultaneous schlieren (top) and OH* chemiluminescence (bottom) images showing multi-jet ignition process in the main chamber for H$_2$/air at $\phi = 0.4$ using the angled multi-nozzle plate with four different spark positions, (**a**) spark 1, (**b**) spark 2, (**c**) spark 3, (**d**) spark 4

The lean flammability limit resulting from the angled multi-jets is shown in Fig. 6.12. The lean flammability limit of angled multi-jets is nearly identical compared to straight multi-jets. For all four spark positions, the flammability limit

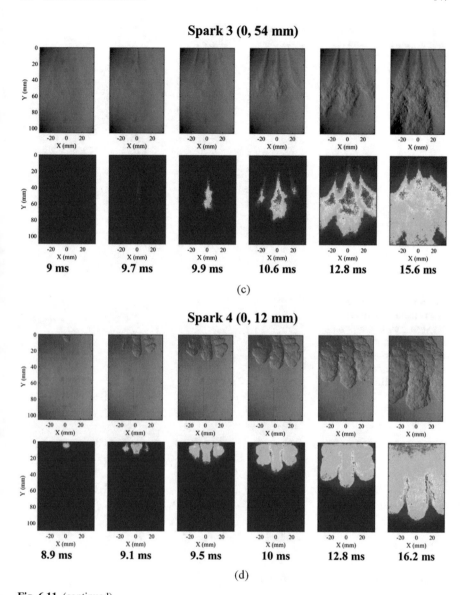

Fig. 6.11 (continued)

ranged from $\phi \approx 0.29$ to 0.31. The ignition delay and ignition probability of the angled multi-jets were closely comparable to the ignition delays of straight multi-jet. Similar to the straight multi-jets, ignition delay of the angled multi-jets decreased as the spark position becomes closer to the nozzle entrance.

Fig. 6.12 Comparison of lean flammability limit of angled multi-jets with straight multi-jets

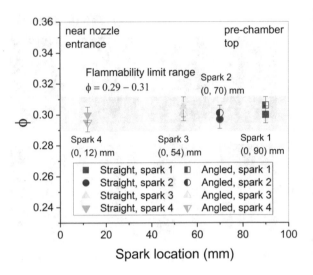

Fig. 6.13 Comparison of main chamber burn time between various single jet and multi-jets. "S" and "A" denote the straight and angled nozzle, respectively, "s" denotes different spark locations

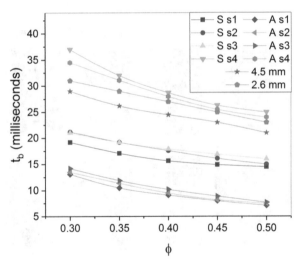

6.3.3.2 Main Combustor Burn Time

A faster burn rate would help to consume the main chamber mixture faster. Depending on the engine speed and operating conditions, a smaller burn time is always preferable during engine operation. However, too fast of a burn rate can lead to engine knock. Figure 6.13 compares the main chamber burn time for single jet and multi-jets for different main chamber equivalence ratios. We found that angled multi-jets produced the smallest burn times for spark positions 1–3. The next fastest burn times were obtained using straight multi-jets for spark positions 1–3. However, the spark position 4, the closest to the nozzle entrance, resulted in the longest burn time. This is because, for spark position 4, ignition was initiated by laminar or slightly turbulent flame jets. The Reynolds number range of the jet was 2500–8000. Since laminar flame propagation is much slower compared to turbulent flame propagation, spark 4 had the longest burn times. Furthermore, the burn times for single jets with a nozzle diameter of 4.5 mm or 2.6 mm were longer compared to multi-jets. Multiple jets brought higher turbulence inside the main chamber, which was believed the reason why burn times for multi-jet was shorter compared to the single jet [39, 40]. Lastly, the reason that angled multi-jets had shorter burn time compared to straight multi-jets is because in the former, jets did not interact with each other. Due to lesser jet interaction, the active flame surface area was higher of the angled-jet configuration compared to the straight jet configuration. This led to faster flame propagation and shorter burn time.

6.3.4 Numerical Modeling of Flame Propagation in Pre-chamber

Our experimental results showed that depending on the spark location and pre-chamber equivalence ratio, the side jets, the middle jet, all the jets, or even laminar flame jets could initiate the main chamber combustion. The most important question to us was: what is the fundamental physics that govern such ignition pattern? To answer this question, we numerically simulated the pre-chamber combustion process.

The purpose was to simulate the transient flame propagation process inside the pre-chamber for various spark locations to understand how that process affects the main chamber ignition pattern. The computational domain is shown in Fig. 6.14a. We modeled the entire cylindrical 3D pre-chamber. The pre-chamber has dimensions of L (length) $\times$ D (diameter) = 88.9 mm (3.5 inches) $\times$ 19.05 mm (1.5 inches). The dimensions of the multi-nozzle plates connecting the two chambers are already reported in Fig. 6.1b. The entire domain except the boundary layer was discretized using tetrahedron cells. Hexahedron cells were used at the boundary. Half a million cells were used in our computation. Figure 6.14b shows the mesh of the center plane cut section. A mesh independence study was conducted by running the model on two different refined meshes – coarser and finer than the original mesh [37]. A pressure outlet boundary condition was used at the nozzle outlets, while everywhere else wall boundary conditions were applied. The initial wall temperature was constant at 300 K with nonslip boundary condition. At the beginning of the simulation, a spark with an energy of 120 mJ was supplied at the specified spark location shown in Fig. 6.3 to initiate ignition.

Unsteady Reynolds averaged Navier–Stokes (U-RANS) equations coupled with mass, energy, and species conservation equations were solved using the commercial code ANSYS Fluent R15.0 [41]. The Reynolds stress models (RSMs) coupled with detailed H_2/air chemistry [42] (9 species, 21 reactions) were implemented. The

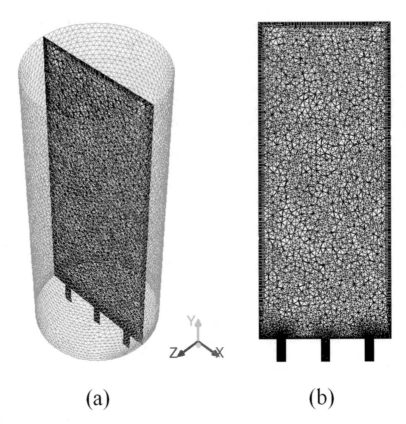

Fig. 6.14 (**a**) 3D pre-chamber domain, (**b**) midplane mesh of the pre-chamber

turbulence-chemistry interaction was modeled using the eddy dissipation concept (EDC) model. The EDC model assumes that reaction occurs in small turbulent structures, called the fine scales. This model has the capability to include detailed chemical reaction mechanisms.

The compressible Navier-Stokes equations were solved using a pressure-based solver in which the pressure and velocity were coupled using the Semi-Implicit Method for Pressure-Linked Equations (SIMPLE) algorithm. At the beginning of the simulation for a few milliseconds, a first-order upwind discretization scheme was used for the convective terms and turbulent quantities to obtain a stable, first-order solution. Once a stable solution was reached, we switched the discretization scheme to third-order Monotone Upstream-Centered Schemes for Conservation Laws (MUSCL) for an accurate solution. However, this higher-order discretization scheme increased computation time significantly. The least squares cell-based gradient calculation scheme, which is known for accuracy and yet computationally less expensive, was chosen over the node-based gradient for the spatial discretization. A second-order discretization scheme was used for pressure. The solution-adaptive mesh refinement feature was used to resolve flame front structure. A dynamic

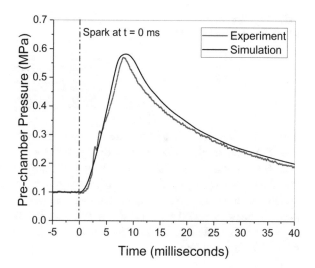

Fig. 6.15 Pre-chamber pressure traces from the model and experiment

adaption of the temperature gradient was implemented to refine the mesh near the flame front or to coarsen it wherever needed. A fixed time step of $t = 10^{-6}$ second was used to resolve the chemical timescale, which was estimated to satisfy the Courant-Friedrichs-Lewy (CFL) condition for numerical stability. To attain a stable solution, we used a Courant number much less than unity. The second-order implicit scheme was used for time integration of each conservation equation. Figure 6.15 compares pre-chamber pressure traces from the model and experiment. The model agrees well with the experiment. Both pressure traces show that after a short ignition delay of 1.2 milliseconds, the pre-chamber was ignited, and the pressure started rising. The peak pressure, which is almost 6 times the initial pressure, occurred at about 9 milliseconds after ignition. Afterward, pressure dropped as the pre-chamber combustion products entered the main chamber.

Before getting into the numerical results, a brief discussion on flame propagation through closed or semi-open ducts is presented below. The flame dynamics inside a small volume of the pre-chamber is much more complex than in a large volume. The challenge arises because of excessive heat losses due to high surface to volume ratio and walls/boundary effects. For example, the heat losses and wall effects during the flame propagation through a narrow tubular-shaped channel may induce instabilities, acoustic waves, vortex flow, and flame acceleration and deceleration [43–46]. For example, a flame propagating in a small tube can undergo a change in shape, from spherical, curved (convex or concave), to the tulip (V-shaped) and cellular fronts [47–49]. The aspect ratio (*L/D* ratio) of the channel is a crucial parameter governing the flame propagation behavior. Several previous studies [50–52] have observed that the hemispherical flame front inverts and a tulip flame occurs in a fully or semi-closed channel when the channel aspect ratio is greater than 2 [43, 45]. Note in these studies, ignition was always achieved at one end of the tube and flame propagated

Table 6.1 The L/D ratio of four different spark locations

Spark location	L/D ($D = 38.1$ mm)
Spark 1 (0, 90 mm)	2.4
Spark 2 (0, 70 mm)	1.9
Spark 3 (0, 54 mm)	1.4
Spark 4 (0, 12 mm)	0.32

toward the other end. Thus, for a pre-chamber, the effective length in the L/D ratio should be the length measured from the ignition location to the other end.

Table 6.1 shows the L/D ratio of the pre-chamber based on four different spark locations. Depending on the spark location, the pre-chamber flame had different length L to travel to enter the main chamber through multi-nozzles. This length L is used to calculated L/D ratio for different spark locations. Spark 1 had L/D ratio higher than 2; spark 2 had L/D ratio 1.9 which was nearly 2. However, spark 3 and 4 had L/D ratios below 2.

Figure 6.16 shows the temperature contours in the pre-chamber at three-time instances near the ignition event taking place in the main chamber and corresponding normalized temperature profiles at the nozzle entrance and nozzle exit. Time $t = 0$ ms signifies ignition is happening in the main chamber. A negative time means before ignition, and a positive time is after the ignition in the main chamber.

Figure 6.16a, b display pre-chamber flame prior to ignition occurring in the main chamber for spark positions 1 and 2. Table 6.1 suggests that for both these spark locations, the effective L/D was higher or close to the critical value of 2. This explains why flame front inversion happened, and the flame front became tulip-shaped. Flame propagated faster near the walls and slowed down at the center. Figure 6.16a shows that 3 milliseconds before main chamber ignition, high-temperature jet flowed through the side nozzles while the center jet temperature was still low. At the instance of ignition, the side jet temperature was higher compared to the middle jet for both spark 1 and spark 2. For spark 2, the temperature difference was much smaller between the side jets and the middle jet. However, the hot products started entering through the side nozzles before it did through the center nozzle. This was the sole reason why side jets initiated the ignition in the main chamber for spark positions 1 and 2 for which the effective L/D was higher than 2.

For spark positions 3 and 4, the L/D ratio was much below the critical ratio of 2 required to achieve flame front inversion. Thus, a convex flame propagated through the center nozzle, and the hot turbulent jet entered into the main combustion chamber firstly through the center nozzle. Thus, the center jet became the initiator of main chamber ignition when L/D ratio was below 2.

For spark position 4, which is the closest to the nozzle entrance, the expanding flame remained laminar when it was entering the center nozzle. Since the flame just started to expand, the pre-chamber sidewall effects were absent, and the flame temperature was much higher compared to the other spark locations. Thus, even after the stretch and heat loss inside the nozzle, the flame did not extinguish; it penetrated into the main chamber in the form of a laminar or slightly turbulent flame jets which then slowly developed into a turbulent flame jet.

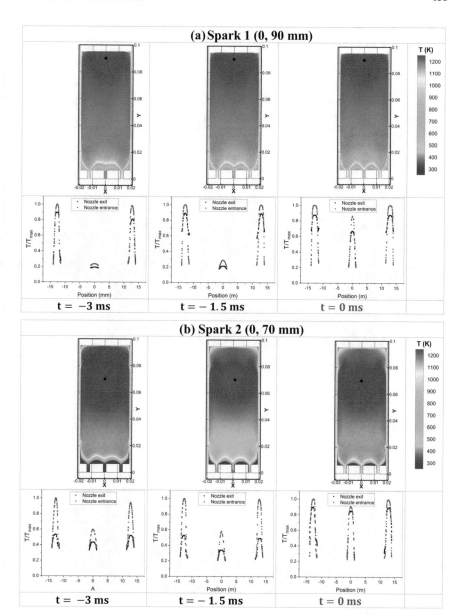

Fig. 6.16 Temperature contours at three-time instances near ignition taking place in the main chamber and corresponding normalize temperature profiles at the nozzle entrance and nozzle exit. The black dot on the temperature contour represents the spark location, (**a**) Spark 1 (0, 90 mm), (**b**) Spark 2 (0, 70 mm), (**c**) Spark 3 (0, 54 mm), (**d**) Spark 4 (0, 12 mm)

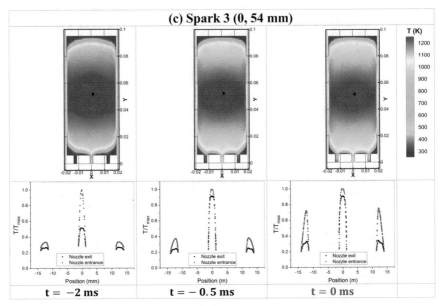

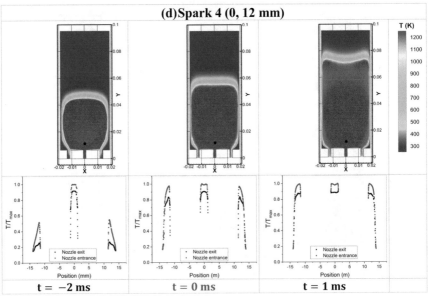

Fig. 6.16 (continued)

6.4 Conclusions

This chapter examines the ignition characteristics of ultra-lean premixed H_2/air by multiple hot turbulent jets in a dual combustion chamber system. The major findings are listed below.

1. Compared to single jet, multiple jets (straight or angled) did not extend the lean flammability limit of H_2/air, given both systems have the same total nozzle area. However, the ignition probability improved significantly near the lean flammability limit by using multiple jets. This is due to the cumulative ignition probability which is higher for multiple hot jet ignition compared to single jets.
2. For multiple jets, the spark location and the fuel/air equivalence ratio inside the pre-chamber have a deterministic effect on the ignition pattern in the main chamber. If the pre-chamber spark is located such that the effective length to diameter ratio exceeds the critical L/D ratio of 2, the side jets ignite the main chamber first. Otherwise, the middle jet or all the jets ignite the main chamber depending on the pre-chamber equivalence ratio. For a specific spark location, a richer H_2/air mixture tended to increase main chamber ignition by all jets. Overall, the effect of pre-chamber equivalence ratio and the spark location was strongly coupled.
3. The numerical simulation results show that the flame shape inside the pre-chamber when approaching the nozzles determined the ignition pattern, e.g., ignition was started by the side jets or the middle jet. For pre-chambers having an effective L/D over 2, flame front inverted and became a tulip-shaped flame, promoting ignition by the side jets.
4. Both experiments and simulations show that if the spark location is too close to the nozzle entrance, ignition was initiated by laminar or slightly turbulent flame jets.
5. Lastly, the main chamber burn rate increased with multi-jets compared to a single jet. Moreover, angled multi-jets increased the burn rate even more compared to straight multi-jets by enhancing turbulence and mixing inside the main combustion chamber.
6. In conclusion, among all the multi-jet configurations, the angled-jet configuration and the spark located at the farthermost location from the nozzle entrance had a superior performance over the others. Even though ignition delay increased slightly due to higher length traveled by the pre-chamber flame before entering into the main chamber, a higher spark location had a better ignition probability and shortest main combustor burn time. This is because at higher spark location, pre-chamber was entirely burned out and that creates a greater pressure difference and a stronger turbulent jet, which promoted mixing and faster combustion in the main chamber.

References

1. Goossak, L.A.: Method of prechamber-torch ignition in internal combustion engines, USPTO, Editor (1966)
2. Gussak, L.A.: The role of chemical activity and turbulence intensity in prechamber-torch organization of combustion of a stationary flow of a fuel-air mixture. In: International Congress & Exposition, Detroit (1983)
3. Gussak, L., Karpov, V., Tikhonov, Y.: The Application of Lag-Process in Prechamber Engines. SAE Technical Paper 790692 (1979)
4. Gussak, L.A., Turkish, M., Siegla, D.: High Chemical Activity of Incomplete Combustion Products and a Method of prechamber Torch Ignition for Avalanche Activation of Combustion in Internal Combustion Engines. SAE Technical Paper 750890 (1975)
5. Oppenheim, A.K.: Quest for controlled combustion engines. In: International Congress and Exposition, Detroit (1988)
6. Oppenheim, A., et al.: Jet Ignition of an Ultra-Lean Mixture. SAE Technical Paper 780637 (1978)
7. Wolfhard, H.G.: The ignition of combustible mixtures by hot gases. J. Jet Propuls. 28(12), 798–804 (1958)
8. Murase, E., et al.: Initiation of combustion in lean mixtures by flame jets. Combust. Sci. Technol. 113(1), 167–177 (1996)
9. Ghoniem, A.F., Oppenheim, A.K., Chen, D.Y.: Experimental and Theoretical Study of Combustion Jet Ignition. California University, Berkeley, Report number: NASA-CR-168139 - DOE/NASA/0131-1 (1983)
10. Pitt, P.L., Ridley, J.D., Clemilnts, R.M.: An ignition system for ultra lean mixtures. Combust. Sci. Technol. 35(5–6), 277–285 (2007)
11. Yamaguchi, S., Ohiwa, N., Hasegawa, T.: Ignition and burning process in a divided chamber bomb. Combust. Flame. 59(2), 177–187 (1985)
12. Elhsnawi, M., Teodorczyk, A.: Studies of mixing and ignition in hydrogen-oxygen mixture with hot inert gas injection. In: Proceedings of the European Combustion Meeting, Warszawa: Warsaw University of Technology ITC, Nowowiejska (2005)
13. Sadanandan, R., et al.: Detailed investigation of ignition by hot gas jets. Proc. Combust. Inst. 31 (1), 719–726 (2007)
14. Toulson, E., et al.: Visualization of propane and natural gas spark ignition and turbulent jet ignition combustion. SAE Int. J. Engines. 5(4), 1821–1835 (2012)
15. Toulson, E., Watson, H., Attard, W.: Gas Assisted Jet Ignition of Ultra-Lean LPG in a Spark Ignition Engine. SAE Technical Paper 2009-01-0506 (2009)
16. Gholamisheeri, M., Wichman, I.S., Toulson, E.: A study of the turbulent jet flow field in a methane fueled turbulent jet ignition (TJI) system. Combust. Flame. 183, 194–206 (2017)
17. Attard, W.P., et al.: A New Combustion System Achieving High Drive Cycle Fuel Economy Improvements in a Modern Vehicle Powertrain. SAE Technical Paper 2011-01-0664 (2011)
18. Perera, I., Wijeyakulasuriya, S., Nalim, R.: Hot combustion torch jet ignition delay time for ethylene-air mixtures. In: 49th AIAA Aerospace Sciences Meeting including the New Horizons Forum and Aerospace Exposition Orlando, Florida., https://doi.org/10.2514/6.2011-95 (2011)
19. Carpio, J., et al.: Critical radius for hot-jet ignition of hydrogen–air mixtures. Int. J. Hydrog. Energy. 38(7), 3105–3109 (2013)
20. Shah, A., Tunestal, P., Johansson, B.: Effect of pre-chamber volume and nozzle diameter on pre-chamber ignition in heavy duty natural gas engines, SAE Technical Paper 2015-01-0867. 1 (2015) https://doi.org/10.4271/2015-01-0867
21. Karimi, A., Rajagopal, M., Nalim, R.: Traversing hot-jet ignition in a constant-volume combustor. J. Eng. Gas Turbines Power. 136(4), 041506 (2013)
22. Thelen, B.C., Toulson, E.: A Computational Study of the Effects of Spark Location on the Performance of a Turbulent Jet Ignition System. SAE Technical Paper 2016-01-0608 (2016)

23. Biswas, S., et al.: On ignition mechanisms of premixed CH4/air and H2/air using a hot turbulent jet generated by pre-chamber combustion. Appl. Therm. Eng. **106**, 925–937 (2016)
24. Biswas, S., Qiao, L.: Prechamber hot jet ignition of ultra-lean H2/air mixtures: effect of supersonic jets and combustion instability. SAE Int. J. Engines. **9**(3), 1584–1592 (2016)
25. Attard, W., et al.: Spark Ignition and Pre-Chamber Turbulent Jet Ignition Combustion Visualization. SAE Technical Paper 2012-01-0823 (2012)
26. Zhang, Q., et al.: Experimental and numerical study of jet controlled compression ignition on combustion phasing control in diesel premixed compression ignition systems. Energies. **7**(7), 4519–4531 (2014)
27. Gentz, G., et al.: A study of the influence of orifice diameter on a turbulent jet ignition system through combustion visualization and performance characterization in a rapid compression machine. Appl. Therm. Eng. **81**, 399–411 (2015)
28. Latsch, R., Schlembach, H.: Externally ignited internal combustion engine. Robert Bosch Gmbh: US 4218992 A (1980)
29. Hideg, L., Ernest, R.P.: Internal combustion engine control system. Ford Motor Company: US4060058 A (1977)
30. Durling, H.E., Johnston, R.P., Polikarpus, K.K.: Torch jet spark plug. General Motors Corporation: US 5421300 A (1995)
31. Anderson, A.C.: Walled precombustion chamber unit. Dresser-Rand Company: US 5533476 A (1996)
32. Nerheim, L.M.: Prechamber for a gas engine. Rolls-Royce Marine: US 20100132660 A1 (2011)
33. Attard, W.: Turbulent jet ignition pre-chamber combustion system for spark ignition engines. MAHLE Powertrain LLC: US 20120103302 A1 (2012)
34. Chiera, D., Hampson, G.J., Polley, N.: Multi-chamber igniter. Woodward, Inc.: US 8839762 B1 (2014)
35. Herold, H., et al.: Pre-combustion chamber tip. Caterpillar Motoren Gmbh & Co. Kg: US 8813716 B2 (2014)
36. Oppenheim, A.K., Stewart, H.E., Hom, K.: Pulsed jet combustion generator for premixed charge engines. The University of California: US 4926818 A (1990)
37. Biswas, S., Qiao, L.: A numerical investigation of ignition of ultra-lean premixed H2/air mixtures by pre-chamber supersonic hot jet. SAE Int. J. Engines. **10**(5), 2231–2247 (2017)
38. Zhao, F., Lai M., Harrington, D.: A Review of Mixture Preparation and Combustion Control Strategies for Spark-Ignited Direct-Injection Gasoline Engines. SAE Technical Paper 970627 (1997)
39. Krutka, H.M., Shambaugh, R.L., Papavassiliou, D.V.: Analysis of multiple jets in the Schwarz melt-blowing die using computational fluid dynamics. Ind. Eng. Chem. Res. **44**(3), 8922–8932 (2005)
40. Peterson, S.D.: Experimental investigation of multiple jets-in-crossflow. Purdue University, West Lafayette 700 (2001)
41. ANSYS: ANSYS Fluent Academic Research, Release 15.0 (2015)
42. Connaire, M.O., et al.: A comprehensive modeling study of hydrogen oxidation. In. J. Chem. Kinet. **36**(11), 603–622 (2004)
43. Ponizy, B., Claverie, A., Veyssière, B.: Tulip flame – the mechanism of flame front inversion. Combust. Flame. **161**(12), 3051–3062 (2014)
44. Xiao, H., et al.: An experimental study of distorted tulip flame formation in a closed duct. Combust. Flame. **160**(9), 1725–1728 (2013)
45. Dunn-Rankin, D., Sawyer, R.F.: Tulip flames: changes in shape of premixed flames propagating in closed tubes. Exp. Fluids. **24**(2), 130–140 (1998)
46. Clanet, C., Searby, G.: On the "tulip flame" phenomenon. Combust. Flame. **105**(1), 225–238 (1996)
47. Xiao, H., et al.: Dynamics of premixed hydrogen/air flame in a closed combustion vessel. Int. J. Hydrog. Energy. **38**(29), 12856–12864 (2013)

48. Xiao, H., et al.: Experimental study on the behaviors and shape changes of premixed hydrogen–air flames propagating in horizontal duct. Int. J. Hydrog. Energy. **36**(10), 6325–6336 (2011)
49. Gonzalez, M., Borghi, R., Saouab, A.: Interaction of a flame front with its self-generated flow in an enclosure; the tulip flame phenomenon. Combust. Flame. **88**(2), 201–220 (1992)
50. Najim, Y.M., Mueller, N., Wichman, I.S.: On premixed flame propagation in a curved constant volume channel. Combust. Flame. **162**(10), 3980–3990 (2015)
51. Magina, N., et al.: Propagation, dissipation, and dispersion of disturbances on harmonically forced, non-premixed flames. Proc. Combust. Inst. **35**(1), 1097–1105 (2015)
52. Hariharan, A., Wichman, I.S.: Premixed flame propagation and morphology in a constant volume combustion chamber. Combust. Sci. Technol. **186**(8), 1025–1040 (2014)

Chapter 7
Impinging Jet Ignition

Contents

7.1 Introduction

Turbulent jet ignition can reliably be used to ignite an ultra-lean fuel/air mixture as illustrated in previous chapters. This ignition technique can be utilized in various applications ranging from pulse detonation engines, wave rotor combustor explosions, to supersonic combustors and natural gas engines. Compared to a conventional spark plug, the hot jet has a much larger surface area leading to multiple ignition sites on its surface which can enhance the probability of successful ignition and cause faster flame propagation and heat release. In short, turbulent jet ignition has many advantages over conventional ignition system.

However, the wall effect will play a key role for the hot turbulent jet ignition used in the engine condition. Inside the small squeeze volume of the engine combustor,

© Springer International Publishing AG, part of Springer Nature 2018   159
S. Biswas, *Physics of Turbulent Jet Ignition*, Springer Theses,
https://doi.org/10.1007/978-3-319-76243-2_7

the effect of confinement and wall effect becomes predominant. The hot jet issued from the pre-chamber may impinge onto the surface of the piston head or the wall of the main engine during the cycle. This is particularly true when multiple jets generated from small pre-chamber nozzle diameters are utilized. Ignition characteristics such as ignition probability, ignition location, and the shape of the ignition kernel might change due to impingement. Thus, we need to understand the impact of impingement on ignition, e.g., piston motion (e.g., the location of the surface) and piston head design (e.g., angle of the surface and geometry of the surface, etc.). This motivated us to study ignition by jet impingement.

The goal of the present study is to examine the ignition behavior of a turbulent hot jet impinging onto a surface. In the experiment, the higher pressure resulting from pre-chamber combustion pushed the combustion products into the main chamber connected by a small diameter nozzle (1.5–3 mm) in the form of a hot turbulent jet, which then impinged on the flat plate and ignited the ultra-lean premixed H_2/air in the main chamber. The distance between the plate center and the nozzle, as well as the plate angle, was varied to understand their effects on ignition.

7.2 Literature Review

Heating of materials by flame impingement has been used for many years in material and processing industries [1–4]. The heat transfer and turbulence characteristic of a jet carrying a reacting gaseous or liquid fuel can enhance dramatically due to impingement. The energy release from an impinging flame jet varies within the subzones near stagnation region. The heat transfer occurs not only by convection but also by radiation from the flame. Malikov [5] showed that from an impinging flame jet, 60–70% heat transfer is convective in nature, while the rest happens in radiation form. However, the sooting probability increases in impinging flame jet devices due to wall effects. Baukal and Baukal and Gebhart [6, 7] developed several semi-analytic models to predict heat transfer characteristics from impinging flame jet.

While a large number of literature have discussed the physics of impinging liquid and gas jets, heat transfer characteristics of the plate from fundamental fluid dynamics and heat transfer standpoint, however, very little exists on ignition of premixed fuel/air by an impinging hot jet. Tajik [8] numerically investigated the heat transfer and emission characteristics of impinging radial jet reattachment combustion (RJRC) flame. They found that the peak heat flux and the concentrations of NOx and CO emissions increase significantly with the increase in Reynolds number. He also observed, with the increase in the nozzle tip to plate spacing, the peak heat flux and the pressure coefficient decrease. Wang [9] studied the ignition process of a methane diffusion impinging flame. They found two types of flame during the ignition process: premixed like a flame with weak blue color and diffusion flame with yellow-reddish color.

All previous studies have focused on two major areas. Fluid dynamics and heat transfer characteristics of impinging non-reacting gas/liquid jets. However, there is little or no work describing the ignition characteristics by high-speed highly turbulent impinging hot jet. This motivated the authors to investigate the ignition of premixed H_2/air using a hot turbulent jet impinging on a flat plate.

The ignition characteristics of a hot turbulent jet impinging on a flat plate surrounded by an ultra-lean premixed H_2/air was studied experimentally. The hot turbulent jet was generated by burning a small quantity of stoichiometric H_2/air mixture in a separate small volume called the pre-chamber. The higher pressure resulting from pre-chamber combustion pushed the combustion products into the main chamber connected by a small diameter nozzle (1.5–3 mm) in the form of a hot turbulent jet, which then impinged on the flat plate and ignited the ultra-lean premixed H_2/air in the main chamber. Six different plates with varying heights and angles were used. Two important parameters controlling the impinging characteristics of the jet, the impinging distance, and the impinging angle were examined. Simultaneous high-speed schlieren and OH* chemiluminescence imaging were applied to visualize the jet penetration and ignition process inside the main combustion chamber. Results illustrate the existence of two distinct types of ignition mechanisms. If the impinging distance is short and the hot turbulent jet hits the plate with high enough momentum, the temperature increases around the stagnation point and the ignition starts from this impinging region. However, if the impinging distance is long, the hot turbulent jet mixes with the unburned H_2/air in the main chamber and ignites the mixture at the upstream from the plate. For such type of ignition, the impinging plate has no role in main chamber ignition. A lower flammability limit of H_2/air was achieved employing the stagnation point ignition, Infrared imaging of the hot jet revealed the radiation intensity profiles near the stagnation region. Effect of jet momentum was studied by varying the nozzle diameters.

7.3 Experimental Methods

The schematic of the experimental setup is shown in Fig. 7.1a, b. The experimental setup was previously described in Chap. 2, and thus only a brief description is presented here. A small volume stainless steel pre-chamber was mounted on the top of a carbon steel main chamber. The main chamber to pre-chamber volume ratio is 100:1. A stainless steel nozzle plate was placed between the two chambers to separate them. In our current experiment, two nozzles with diameters $D = 3$ mm and 1.5 mm were used. Mixtures in both the chambers were initially kept at room temperature. The stoichiometric H_2/air mixture in the pre-chamber was ignited by an electric spark generated at the top of the pre-chamber. Once the spark ignited the pre-chamber mixture, the combustion products started to enter the main chamber in the form of a hot jet which then impinged on the flat plates kept inside the main chamber and ignited the ultra-lean H_2/air in the main chamber. The lean flammability limit for each nozzle was found by gradually reducing the H_2/air equivalence ratio inside the main chamber until ignition could not occur anymore. Note the H_2/air

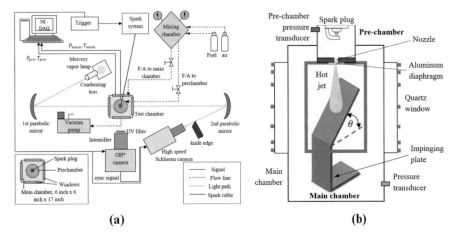

Fig. 7.1 Schematic of (**a**) the experimental setup [10], (**b**) the dual combustion chamber with an impinging plate inside the main chamber

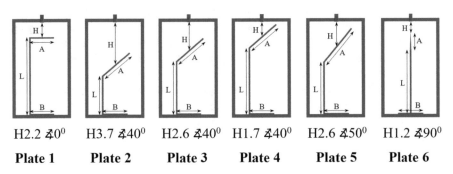

Fig. 7.2 Schematic of the impinging plates inside the main combustion chamber

equivalence ratio of the pre-chamber mixture was fixed at $\phi = 1$ for all cases, whereas the H_2/air equivalence ratio of the main chamber mixture was varied.

Six different stainless steel impinging plates were used in the experiment. The schematic of the impinging plates and plate numberings are shown in Fig. 7.2. In the plate numbering, there are two quantities. The first quantity after "H" denotes impinging height/distance in inch. Impinging height is the normal distance between the nozzle exit and the impinging plate along the nozzle centerline. The second quantity "$\measuredangle$" denotes the impinging angle. We have defined the angle in Fig. 7.1. It is the angle of the impinging plane with the horizontal direction. Using this plate numbering scheme, H2.2 $\measuredangle 0^{\circ}$ denotes an impinging plate 2.2 inches away from the jet exit, and the impinging plane makes an angle zero degree with the horizontal direction.

The plate dimensions are reported in Table 7.1. For all the plates, the base length, width, and thickness were identical. The base length, width, and thickness were 3 inches, 2 inches, and 0.2 inches, respectively. The impinging surface was smooth

Table 7.1 Dimensions of the all the impinging plates

Plate #	Plate name	Impinging surface length, A, inch	Straight section, L, inch	Distance from nozzle exit, D, inch	H/D	θ
1	H2.2 ∡0°	3	14.8	2.2	18.5	0°
2	H3.7 ∡40°	3.9	15	3.7	23.5	40°
3	H2.6 ∡40°	3.9	16	2.6	22	40°
4	H1.7 ∡40°	3.9	17	1.7	14.5	40°
5	H2.6 ∡50°	4.7	14.9	2.6	22	50°
6	H1.2 ∡90°	0	15.8	1.2	10	90°

and free from any irregularities. After every five tests, the surface was thoroughly cleaned to avoid any water deposition which is the only product of lean hydrogen combustion. The two most important geometric parameters of the impinging jet are H/D and impinging angle. They are reported in Table 7.1. The single most crucial factor determining the heat transfer characteristics by impinging jet is impinging height to nozzle diameter ratio or H/D ratio. For a turbulent jet, heat transfer reaches a maximum at the end of the potential core because, after the potential core, turbulence intensity increases. The jet impingement angle can change the heat transfer characteristics. The impinging jet on inclined surface distorts symmetry in the heat transfer contours, generated elliptical isoclines for the Nusselt number. It was observed that if the impinging angle is increased, the Nusselt number decreases [11, 12]. For all the test conditions, unless otherwise stated, 3 mm diameter nozzle was used. To explore the effect of nozzle diameter, a 1.5 mm nozzle was used. Impinging jet ignition results were compared with usual hot turbulent jet ignition cases without any impinging plate present in the main combustion chamber. These cases were denoted as "NP" or no plate condition in legends.

7.3.1 High-Speed Schlieren and OH* Chemiluminescence Imaging

A customized trigger box synchronized with the CDI spark ignition system sent a master trigger to two high-speed cameras for simultaneous schlieren and OH* chemiluminescence imaging. The main chamber was installed with four rectangular (14 cm × 8.9 cm × 1.9 cm) quartz windows (type GE124) on its sides for optical access. One pair of the windows was used for the z-type schlieren system. Another pair was selected for simultaneous OH* chemiluminescence measurements.

The high-speed schlieren technique was utilized to visualize the evolution of the hot jet as well as the ignition process in the main chamber. The system consisted of a

100 W (ARC HAS-150 HP) mercury lamp light source with a condensing lens, two concave parabolic mirrors (15.24 cm diameter, focal length 1.2 m), and a high-speed digital camera (Vision Research Phantom v7). Schlieren images were captured with a resolution of 800 × 720 pixels with a frame rate up to 12,000 fps.

The high-speed OH* chemiluminescence measurement provided a better view of the ignition and flame propagation processes. A high-speed camera (Vision Research Phantom v640), along with video-scope-gated image intensifier (VS4-1845HS) with 105 mm UV lens, were utilized to detect OH* signals at a very narrow band 386 ± 10 nm detection limit. The intensifier was externally synced with the camera via a high-speed relay and acquired images at the same frame rate (up to 12,000 fps) with the Phantom camera. A fixed intensifier setting (gain 65,000 and gate width 20 microseconds, aperture f8) was used all through.

7.3.2 Infrared Imaging

Planar time-dependent radiation intensity measurements of the flame were acquired using an infrared camera (FLIR SC6100) with an InSb detector. The view angle of the camera was aligned perpendicular to the flame axis (50 cm from the burner center to the camera lens) such that the half view angle of the camera is less than 10°. The radiation intensity detected by each pixel of the camera focal plane array can be approximated by a parallel line-of-sight because of the small view angle. The spatial resolution is 0.2×0.2 mm^2 for each pixel. The band-pass filter was used to measure the radiation intensity of H2O (2.58 ± 0.03 μm).

7.4 Results and Discussion

Results and discussion are divided into following subsections. First, we discuss the ignition pattern in the main chamber using impinging jet ignition. The lean flammability limit is examined and discussed in the next section. Then we look at the infrared images just at the prior to ignition in the main chamber to understand the effect of temperature distribution on the ignition mechanism. We discuss the effect of different nozzle diameters on the ignition mechanism. Lastly, we introduce Nusselt number correlations and compare the heat transfer performances from different nozzles.

7.4.1 Flammability Limit and Ignition Delay

We found an interesting result using the impinging hot turbulent jet. The lean flammability limit of H$_2$/air could be extended by impinging hot jet ignition. Without

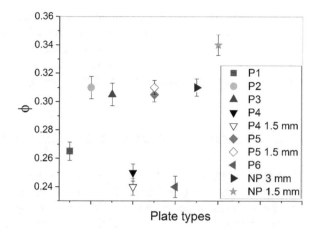

Fig. 7.3 Flammability limit of H$_2$/air using impinging hot turbulent jet ignition. "P" and "NP" denotes plate and no plate inside the main combustion chamber

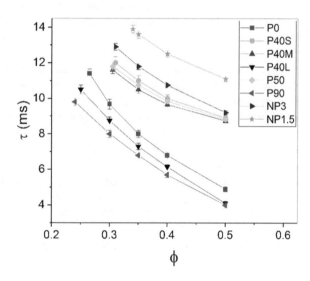

Fig. 7.4 Ignition delay for various impinging and turbulent jet ignition

any impinging plate, the limit was $\phi_{\text{limit}} = 0.31$. Using impinging jets, this limit was extended to $\phi_{\text{limit}} = 0.24$. Figure 7.3 plots the lean flammability limit of different impinging plate and nozzle combinations. Figure 7.3 also compares impinging jet ignition test cases with turbulent jet ignition cases.

Out of six impinging plates we used, only three plates were successful to extend the lean flammability limit of H$_2$/air. Those plates were 1, 4, and 6. Note that the H/D ratio of these three plates was minimum compared to the other plates; they were 18.5, 14.5, and 10, respectively. Additionally, we used different nozzle diameters on plates 4 and 5 to understand how the nozzle diameter affects the ignition mechanism. Figure 7.3 shows 1.5 mm diameter nozzle could extend the limit for plate 4 only. Comparing all the plates, we can conclude that the major factor affecting the lean flammability limit is the impinging distance.

Figure 7.4 shows the ignition delay for different plates. Plates 1, 4, and 6 produce the smallest ignition delay in the main chamber. The reason for shorter ignition delay is hiding behind the impinging jet ignition mechanism. As soon as the hot jet hits the plate with sufficient heat and energy, the thermal energy is released near the impinging location, increasing the local temperature. At the same time, the flow velocity near impinging location is rather small or nearly stagnant. This creates an ideal environment for ignition to take place.

7.4.2 Ignition Visualization in the Main Chamber

High-speed schlieren and OH* imaging enabled us to visualize the ignition inside the main chamber. Figure 7.5a–c show the time sequence of ignition in the main chamber by an impinging jet. OH* marks the onset of ignition in the main chamber. Figure 7.5 shows for plates 1, 4, and 6 the ignition started from the surface of the impinging plates. This suggests that as the hot turbulent jet impinged onto the plate, the kinetic energy of the hot jet was transferred into thermal energy. Thus, the temperature around the stagnation point increased, which led to the ignition of the ultra-lean mixture in the main chamber.

Examining the impinging distance, we found the maximum impinging distance was 2.2 inches that could result in ignition by an impinging hot jet. Anything above 2.2 inches would not cause a reliable ignition by the impinging ignition mechanism. Another interesting observation was the ignition by plate 6, H1.2 ∡90°. As the jet split into two by the sharp plate 6, the jet started to slow down due to boundary effect. The ignition started from the boundary layer on the plate.

However, for plates 2, 3, and 5, the ignition occurred at the upstream of the impinging point. The ignition mechanism is the same as what we previously found that the ignition happens on the lateral sides of the jet before it hits the impinging surface [10]. In this case, the Damköhler number can be used to describe the ignition outcome, which is a result of turbulent mixing between the hot turbulent jet of combustion products and the cold, unburned ambient lean mixture.

7.4.3 Effect of Nozzle Diameter

Figure 7.6 shows the ignition process inside the main chamber for nozzle diameter 1.5 mm. For plate 4 ignition started after the jet impinged on the plate. However, for plate 5, ignition initiated from upstream of the impinging point. The outcome of the 1.5 mm nozzle is same as the 3 mm nozzle. Plate 4 showed impinging jet ignition, while plate 5 resulted in ignition by typical turbulent jet ignition mechanism.

Figure 7.7 shows the effect of nozzle diameters on ignition delay. It is interesting to note that when the ignition did not occur by impingement, the ignition delay remained the same. However, when the ignition occurred through jet impingement, using a smaller nozzle produced a lower ignition delay. This is an interesting observation, and we must understand the impingement physics of the jet which will be discussed in following sections.

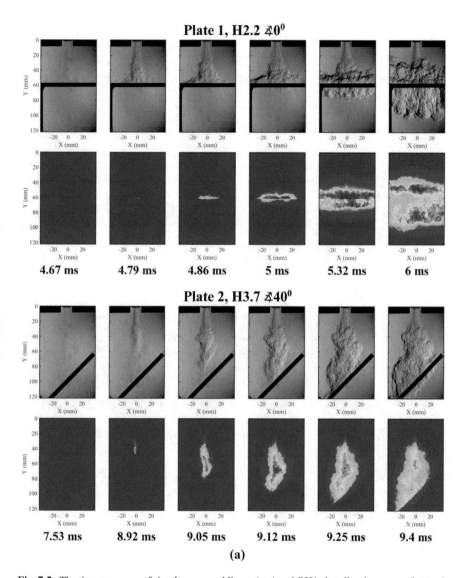

Fig. 7.5 The time sequence of simultaneous schlieren (top) and OH* chemiluminescence (bottom) images showing impinging jet ignition process for H_2/air for impinging plates (**a**) 1 and 2, (**b**) 3 and 4, and (**c**) 5 and 6. $V_{pre-chamber} = 100cc$, $d_{orifice} = 3$ mm, $P_{initial} = 0.1$ MPa, $T_{initial} = 300$ K, $\phi_{pre-chamber} = 1.0$, $\phi_{main\ chamber} = 0.4$

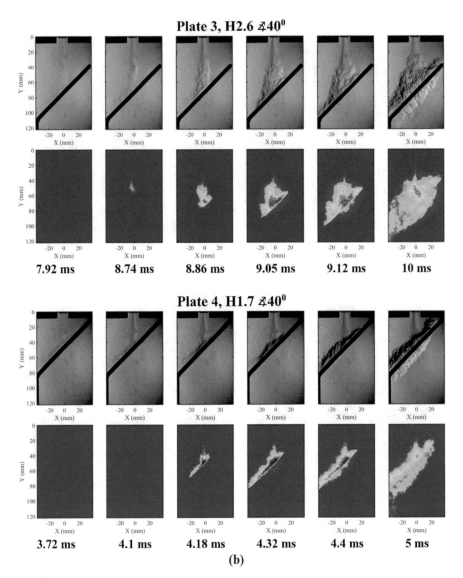

Fig. 7.5 (continued)

7.4.4 Infrared Imaging of Impinging Jets

Figure 7.8 shows simultaneous planar time-dependent radiation intensity measurements and high-speed schlieren imaging to visualize high-temperature zones near the stagnation region. Plates 1, 4, and 6 show an elevated intensity from infrared images

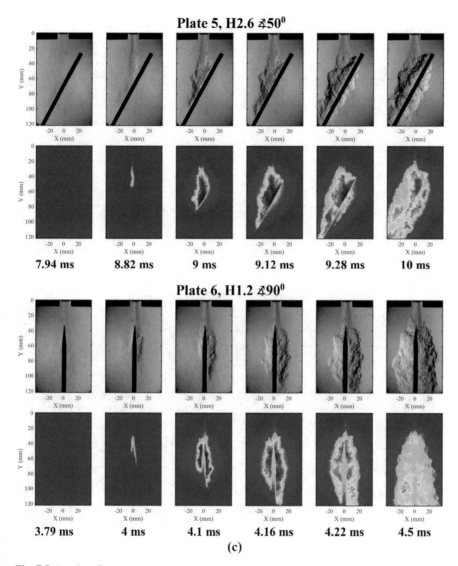

Fig. 7.5 (continued)

near the impinging point. These are the only three plates that could ignite the fuel/air by impinging jet ignition mechanism. Figure 7.9 shows the infrared images of jet impingement from 1.5 mm nozzle diameter. An elevated temperature region was observed near the impinging surface only for plate 4. For plate 5 there is no such elevated temperature region.

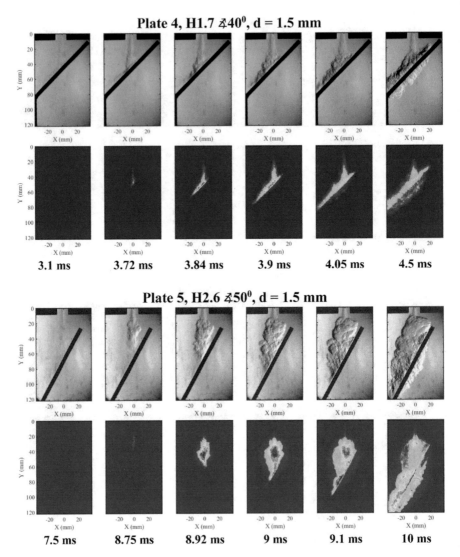

Fig. 7.6 Time sequence of simultaneous schlieren (top) and OH* chemiluminescence (bottom) images showing impinging ignition process for H_2/air for impinging plates 4 and 5 using a nozzle diameter 1.5 mm. $V_{pre-chamber} = 100$ cc, $d_{orifice} = 1.5$ mm, $P_{initial} = 0.1$ MPa, $T_{initial} = 300$ K, $\phi_{pre-chamber} = 1.0$, $\phi_{main\ chamber} = 0.4$

7.4.5 *Radiation Intensity*

Radiation intensity of the impinging region explains better why plates 1, 4, and 6 could ignite via impinging jet ignition when the other plates failed to do so. Figure 7.10b shows the local $\eta - \xi$ coordinate. The radiation intensity 2 mm away from the impinging surface is plotted in Fig. 7.10a.

Fig. 7.7 Effect of nozzle diameters on ignition delay

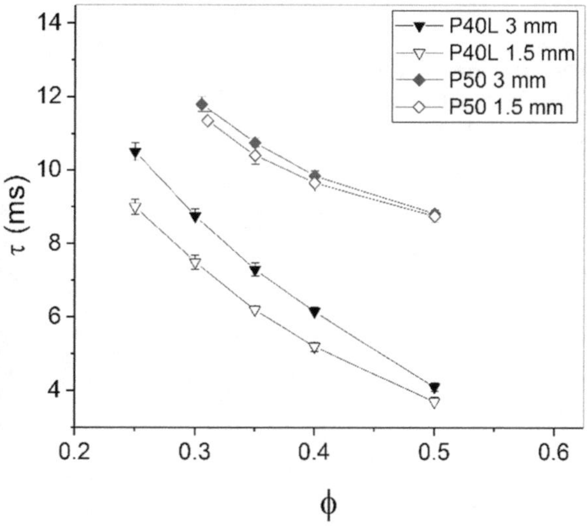

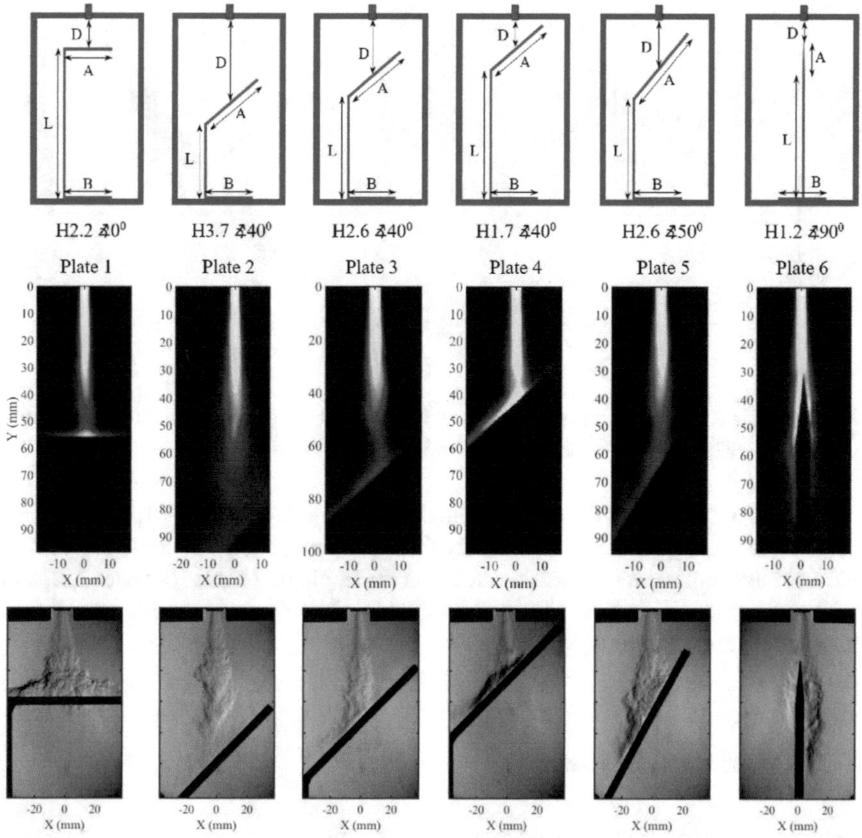

Fig. 7.8 Simultaneous planar time-dependent radiation intensity measurements of the hot imping-ing jet with H_2O (2.58 ± 0.03 μm) band-pass filter and high-speed schlieren imaging reveal the shock structure of supersonic jets. Infrared (top) and schlieren images (bottom) showing the hot turbulent heat impinging on the surface just before ignition in the main chamber

Fig. 7.9 Infrared imaging
of impinging jets generated
from 1.5 mm diameter
nozzle

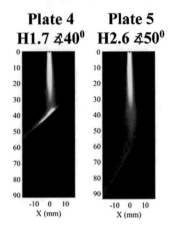

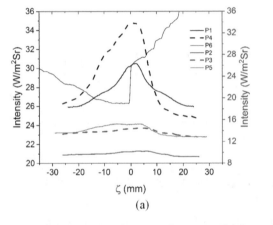

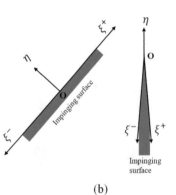

Fig. 7.10 (**a**) Radiation intensity along a parallel line to the impinging surface with an offset distance of 2 mm for different impinging plates using a 3 mm nozzle, (**b**) the local $\eta - \xi$ coordinate

The black line shows the radiation intensity for impinging plates 1, 4, and 6. All of them have a higher radiation intensity than the other plates. This indicates that an elevated temperature region was created near the impinging plane for plates 1, 4, and 6. Next section discusses why and how this elevated temperature region was established for impinging plates.

7.5 Numerical Modeling

Our experimental results showed that depending on the plate location and impingement angle, only plates 1, 4, and 6 could initiate the main chamber combustion. The most important question to us was: what is the fundamental physics that govern impinging jet ignition? To answer this question, we numerically simulated the reacting jet impingement process.

7.5.1 Simulation Domain and Boundary Conditions

The purpose was to simulate the jet impingement inside a non-reacting main chamber for various impinging plates to understand how impingement process affects the mainchamber ignition pattern. The computational domain is shown in Fig. 7.11. Due to symmetry, we modeled half of the 3D cylindrical pre-chamber, nozzle, and rectangular main chamber with the impinging plate. The pre-chamber and main chamber have the exact same dimensions that were used for experiments. The dimensions of the impinging plates are already reported in Fig. 7.2. The entire domain except the boundary layer was discretized using tetrahedron cells. Hexahedron cells were used at the boundary. Five million cells were used in our computation. A mesh independence study was conducted by running the model on two different refined meshes – coarser and finer than the original mesh [13]. A pressure outlet boundary condition was used at the nozzle outlets, while everywhere else wall boundary conditions were applied. The initial wall temperature was constant at 300 K with nonslip boundary condition. At the beginning of the simulation, a spark with an energy of 120 mJ was supplied at the specified spark location shown in Fig. 7.1b to initiate ignition.

7.5.2 Numerical Details

Unsteady Reynolds averaged Navier–Stokes (U-RANS) equations coupled with mass, energy, and species conservation equations were solved using the commercial code ANSYS Fluent R15.0 [14]. The Reynolds stress models (RSMs) coupled with

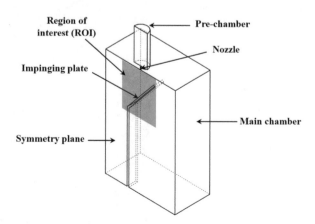

Fig. 7.11 Simulation domain for impinging jets

detailed H$_2$/air chemistry [15] (9 species, 21 reactions) were implemented. The turbulence-chemistry interaction was modeled using the eddy dissipation concept (EDC) model. The EDC model assumes that reaction occurs in small turbulent structures, called the fine scales. This model has the capability to include detailed chemical reaction mechanisms.

The compressible Navier-Stokes equations were solved using a pressure-based solver in which the pressure and velocity were coupled using the Semi-Implicit Method for Pressure-Linked Equations (SIMPLE) algorithm. At the beginning of the simulation for a few milliseconds, a first-order upwind discretization scheme was used for the convective terms and turbulent quantities to obtain a stable, first-order solution. Once a stable solution was reached, we switched the discretization scheme to third-order Monotone Upstream-Centered Schemes for Conservation Laws (MUSCL) for an accurate solution. However, this higher-order discretization scheme increased computation time significantly. The least squares cell-based gradient calculation scheme, which is known for accuracy and yet computationally less expensive, was chosen over the node-based gradient for the spatial discretization. A second-order discretization scheme was used for pressure. The solution-adaptive mesh refinement feature was used to resolve flame front structure. A dynamic adaption of the temperature gradient was implemented to refine the mesh near the flame front or to coarsen it wherever needed. A fixed time step of $t = 10^{-6}$ s was used to resolve the chemical timescale, which was estimated to satisfy the Courant-Friedrichs-Lewy (CFL) condition for numerical stability. To attain a stable solution, we used a Courant number much less than unity. The second-order implicit scheme was used for time integration of each conservation equation. Figure 7.12 compares pre-chamber pressure traces from the model and experiment. The model agrees well with the experiment. Both pressure traces show that after a short ignition delay of 1.2 milliseconds, the pre-chamber was ignited, and the pressure started rising. The peak pressure, which is almost 6 times the initial pressure, occurred at about 9 milliseconds

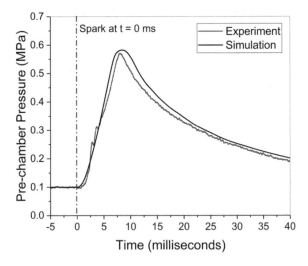

Fig. 7.12 Pre-chamber pressure traces from the model and experiment

after ignition. Afterward, pressure dropped as the pre-chamber combustion products entered the main chamber.

7.5.3 Numerical Results of Jet Impingement

Numerical results for six different impinging geometries are presented in Fig. 7.13 just before ignition occurs in the main chamber. Quantitative comparison of velocity, pressure, temperature, and fuel mass fraction ζ of different impinging plate geometry were performed to explain the superior performance of plates 1, 4, and 6. The fuel mass fraction has been defined in Chap. 4, Sect. 4.4.2.3.4.

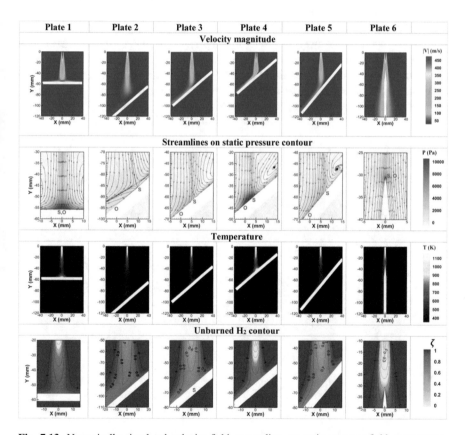

Fig. 7.13 Numerically simulated velocity field, streamline on static pressure field, temperature field, and fuel mass fraction ratio, ζ just before ignition in the main chamber for H_2/air for impinging plates 1–6. $V_{pre-chamber} = 100$ cc, $d_{orifice} = 3$ mm, $P_{initial} = 0.1$ MPa, $T_{initial} = 300$ K, $\phi_{pre-chamber} = 1.0$, $\phi_{main\ chamber} = 0.4$

The ignition delay for impinging jet ignition is smaller compared to typical jet ignition. For example, at equivalence ratio $\phi = 0.4$ ignition delay for impinging jet ignition is almost half, ~4–5 ms compared to typical jet ignition, ~8–12 ms. Since the jet velocity is driven by the pressure different in pre-chamber and main chamber, at lower ignition delay time, the pressure difference remains high and the jet velocity is higher at the ignition condition for impinging jets as evident from Fig. 7.13. Figure 7.13 also shows that the static pressure increases at the stagnation point for plates 1, 4, and 6. Also, the stagnation point moves upward on the plate with increase in plate angle and distance from nozzle exit. Interestingly main chamber ignition always initiated from the region between "OS" as shown in Fig. 7.13. The significance of "OS" will be discussed in detail in the next section. Figure 7.13 shows the temperature profiles of the impinging jet just before ignition in the main chamber. Temperature rise can be observed near the stagnation region for plates 1, 4, and 6. The fuel mass fraction shows the mixing behavior for different plates. For plates 1, 4, and 6, the mixing started to occur near the impinging region. For other places, the jet gets well-mixed even before it impinges on the plate. The key factor that governs the mixing is the distance of the impinging surface from the nozzle exit. These numerical results match excellently with our experimental observations.

Figure 7.14 plots the numerically calculated static pressure, static temperature, turbulent dissipation, velocity, and vorticity magnitude at three distinct locations,

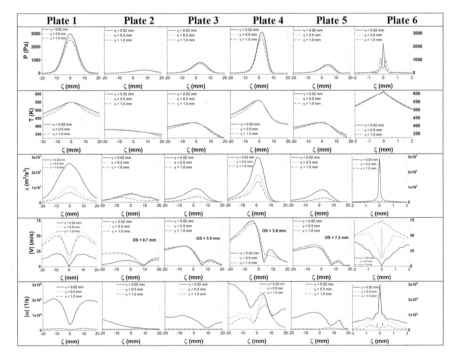

Fig. 7.14 Numerically calculated static pressure, static temperature, turbulent dissipation, velocity, and vorticity magnitude at three distinct locations, $\eta = 0.04$, 0.5 and 1.0 mm from the impinging surface

$\eta = 0.04$, 0.5 and 1.0 mm from the impinging surface for six plates. For plates 1, 4, and 6, the static pressure and temperature increases significantly near the stagnation region compared to other three plates. Around the similar region, the turbulent dissipation and the vorticity magnitude increase as well. Thus, it shows that for plates 1, 4, and 6 the turbulent jet has higher local temperature and superior mixing characteristics. This led to ignition by impingement for plates 1, 4, and 6. Another interesting fact is the behavior of the distance between geometric center and the stagnation point, "OS". For identical impinging angle, "OS" decreases with decreasing impinging distance. A larger "OS" signifies that the static pressure and temperature will be distributed over a larger area. This could potentially decrease the jet temperature near impinging surface and lower the chance of ignition.

7.5.4 Impinging Jet Dynamics

The impinging jet helps to enhance mixing. Impinging introduces higher turbulence in the surrounding mixture. Impingement jet can produce heat transfer that is up to three to four times higher compared to forced convection, by virtue of thin impinging boundary layer. However, the jet lasts for in the order of milliseconds (<25 milliseconds) before it ignites the main chamber. Within such brief period, heat transfer from the jet to plate is negligible. To evaluate how much heat could transfer from the impinging jet to the plate, we calculated the Nusselt number [16],

$$\mathrm{Nu} = \frac{hL}{k_f} \tag{7.1}$$

where h is the convective heat transfer coefficient between the jet and the plate, and k_f is the thermal conductivity of the hot turbulent jet, L is the characteristic length scale. For a round jet, the characteristic length scale is the nozzle diameter. To calculate the Nusselt number, we used the following correlation developed by Beitelmal [17].

$$\mathrm{Nu}_{maximum} = 0.09821 Re^{0.7}(1 + 0.365 \, \sin\theta)\left(1 - 0.0248\frac{H}{D}\right) \tag{7.2}$$

As can be seen in Eq. 7.2, the Nusselt number depends on the Reynolds number, the impinging angle, θ, and H/D ratio. We found the range of Nusselt number for all our plates is in the order of hundred. Since $\mathrm{Nu} \gg 1$, we safely assume the heat transfer from the hot impinging jet to the plate was negligible.

To understand how impinging jets could extend lean flammability limit, we need to look closely at the fluid dynamics, heat and mass transfer in and around the impinging region. Figure 7.15 shows the schematic of a jet impinging on an inclined surface. For a plate placed normal to the impinging direction such as plate 1, the stagnation point "S" coincides with "O," the geometric center of the impinging plate as shown in Fig. 7.2. As the impinging angle increases, the stagnation point starts to

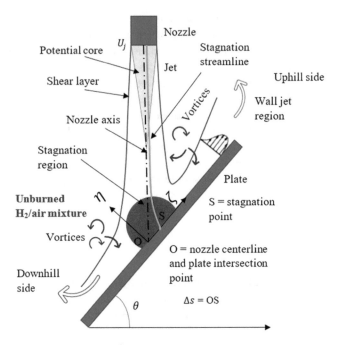

Fig. 7.15 Displacement of the stagnation point in angled impinging jet shown schematically [2]

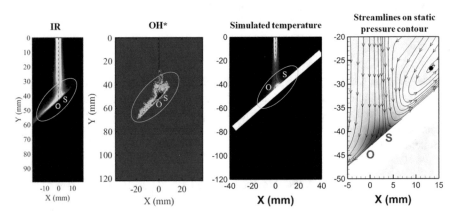

Fig. 7.16 Velocity, temperature, and static pressure field around the stagnation point

move upward [2]. Figure 7.16 support this observation, which shows a hot turbulent jet of Reynolds number 27,450 impinged on a plate with an inclination angle of 40° and the stagnation point shifted upward on the plate. The shifting can be noticed from the higher intensity of the IR image occurred at the upstream of the geometric center "O" as well as from numerical results. OH* image shows the location from where the ignition initiated in the main chamber. Our numerical results show that the static pressure increase is maximum between "O" and "S." Interestingly, ignition in the main chamber started from the region between "O" and "S."

The complex fluid dynamics around the impinging region holds the key for the lean limit extension. As shown schematically in Fig. 7.15, when the hot turbulent jet impinges on the plate, the jet comes to a standstill at the stagnation point. The jet velocity in the stagnation region slows down considerably. Due to a sudden change in the flow direction, the vorticity in the impinging region increases. After a while, the turbulent jet becomes reattached to the plate in the wall jet region. However, between the stagnation and wall jet region, the jet narrows down and the mixing increases. For a turbulent jet with Reynolds number higher than 15,000, local mixing pockets arise in this narrow wall jet region. Through this region, the unburned fuel/air can mix well with the turbulent jet [1]. All the kinetic energy in the high velocity hot turbulent jet converts into thermal energy at the stagnation point. Thus, the dynamic temperature rise occurs at the stagnation point "S." Due to enhanced mixing, the unburned fuel/air entrains in the hot jet through impinging region. If the unburned fuel/air arrives near to this stagnation region, due to high-temperature condition at the stagnation region, ignition occurs.

We have noticed ignition by an impinging jet always started from the zone, "OS." Ignition by impingement directly related to two key parameters, H/D ratio, and the impinging angle. As we increase the distance between nozzle exit to the plate, the hot jet velocity delays and when finally, it impinges on the plate, remaining very little kinetic energy converts into thermal energy. Thus, ignition occurs by typical jet ignition mechanism.

As we increase the impinging angle, the distance "OS" increases. Thus, the thermal energy spreads over a larger "OS" making the overall temperature increase small. Thus, excessive impinging angle did not result in the ignition as shown in Fig. 7.5. As shown in Fig. 7.16 at the uphill side of the impingement region, the flow is less turbulent but contains more heat since heat dissipation is slower. At the downhill side of the impingement, the flow loses heat quickly due to higher turbulence and faster mixing [1]. Thus, we need to optimize the distance, "OS." A balance between H/D ratio and the impinging angle facilitates ignition by impingement.

7.6 Conclusions

Present study experimentally investigated the ignition characteristics of ultra-lean H_2/air mixtures using an impinging hot turbulent jet generated by pre-chamber combustion. Our major findings are summarized below.

Impinging jet ignition can extend the flammability limit of the lean H_2/air mixture in the main chamber, as compared to typical jet ignition. This is because impinging jet ignition creates a high-temperature, stagnation zone near the impinging point and ignition starts from this zone. Enhanced mixing due to increase in vorticity near the impinging region allows surrounding unburned fuel/air to penetrate and enter into the stagnation region. A high dynamics temperature rise in the stagnation region facilitates ignition. This is the reason why impinging jet could extend the lean flammability limit of H_2/air. Furthermore, impinging jet ignition decreases the ignition delay in the main chamber.

Out of six different impinging plates, only three were able to ignite the fuel/air mixture by the impinging jet ignition mechanism. The dominating factor is the impinging distance, which determines the probability of impinging jet ignition.

The dominating factor in the impinging jet ignition is the impinging distance and impinging angle, which determines the probability of impinging jet ignition. A higher impinging distance does not result in impinging jet ignition. Rather, ignition occurs by typical turbulent jet ignition mechanism. With the increase in the impinging angle, the stagnation point shifts away from the jet centerline and thus the probability of impinging jet ignition decreases.

References

1. Viskanta, R.: Heat transfer to impinging isothermal gas and flame jets. Exp. Thermal Fluid Sci. **6**(2), 111–134 (1993)
2. Martin, H.: Heat and mass transfer between impinging gas jets and solid surfaces. Adv. Heat Tran. **13**, 1–60 (1977)
3. Donaldson, C.D., Snedeker, R.S.: A study of free jet impingement. Part 1. Mean properties of free and impinging jets. J. Fluid Mech. **45**(2), 281–319 (1971)
4. Jambunathan, K., et al.: A review of heat transfer data for single circular jet impingement. Int. J. Heat Fluid Flow. **13**(2), 106–115 (1992)
5. Malikov, G.K., et al.: Direct flame impingement heating for rapid thermal materials processing. Int. J. Heat Mass Transf. **44**(9), 1751–1758 (2001)
6. Baukal, C.E., Gebhart, B.: A review of empirical flame impingement heat transfer correlations. Int. J. Heat Fluid Flow. **17**(4), 386–396 (1996)
7. Baukal, C.E., Gebhart, B.: A review of semi-analytical solutions for flame impingement heat transfer. Int. J. Heat Mass Transf. **39**(14), 2989–3002 (1996)
8. Tajik, A.R., Hindasageri, V.: A numerical investigation on heat transfer and emissions characteristics of impinging radial jet reattachment combustion (RJRC) flame. Appl. Therm. Eng. **89**, 534–544 (2015)
9. Wang, Q., Zhao, C.Y., Zhang, Y.: Time-resolved 3D investigation of the ignition process of a methane diffusion impinging flame. Exp. Thermal Fluid Sci. **62**, 78–84 (2015)
10. Biswas, S., et al.: On ignition mechanisms of premixed CH4/air and H2/air using a hot turbulent jet generated by pre-chamber combustion. Appl. Therm. Eng. **106**, 925–937 (2016)
11. Tawfek, A.A.: Heat transfer studies of the oblique impingement of round jets upon a curved surface. Heat Mass Transf. **38**(6), 467–475 (2002)
12. Sparrow, E.M., Lovell, B.J.: Heat transfer characteristics of an obliquely impinging circular jet. J. Heat Transf. **102**(2), 202 (1980)
13. Biswas, S., Qiao, L.: A numerical investigation of ignition of ultra-lean premixed H2/air mixtures by pre-chamber supersonic hot jet. SAE Int. J. Engines. **10**(5), 2231–2247 (2017)
14. ANSYS.: ANSYS Fluent Academic Research, Release 15.0 (2015)
15. Connaire, M.O., Curran, H.J., Simmie, J.M., Pitz, W.J., Westbrook, C.K.: A comprehensive modeling study of hydrogen oxidation. In. J. Chem. Kinet. **36**(11), 603–622 (2004)
16. Bergman, T.L., Incropera, F.P.: Fundamentals of Heat and Mass Transfer, 6th edn, John Wiley, Hoboken (2007)
17. Beitelmal, A.H., Saad, M.A., Patel, C.D.: The effect of inclination on the heat transfer between a flat surface and an impinging two-dimensional air jet. Int. J. Heat Fluid Flow. **21**(2), 156–116 (2000)

Chapter 8
Flame Propagation in Microchannels

Contents

8.1 Introduction

Combustion at small scales (micro- and mesoscales) is gaining increasing attention these days due to the wide spectrum of potential applications in sensors, actuators, portable electronic devices, rovers, robots, unmanned air vehicles, thrusters, industrial heating devices, and, furthermore, heat and mechanical backup power sources for air-conditioning equipment in hybrid vehicles and direct ignition (DI) engines as well [1–3]. Combustion of hydrocarbon fuels is more attractive to manufacturers of miniature power devices because the energy density of hydrocarbons is several times higher than modern batteries [4]. Microscale combustion physics is quite different from those at larger length scales. For example, flame propagation through narrow channels has unique characteristics, e.g., the increasing effects of flame–wall interaction and molecular diffusion [5–10]. In small-scale combustion systems, the surface-to-volume (S/V) ratio is large, which leads to more heat loss and thus causes flame extinction more easily.

© Springer International Publishing AG, part of Springer Nature 2018
S. Biswas, *Physics of Turbulent Jet Ignition*, Springer Theses,
https://doi.org/10.1007/978-3-319-76243-2_8

The present investigation of flame propagation through microchannels was motivated by the previous work on ignition of ultra-lean mixtures by using a turbulent hot jet [11–14]. Pre-chamber hot-jet ignition has been used for heavy-duty natural gas engines, as well as in pulse detonation engines, where a small-diameter orifice (or nozzle) is used to connect a pre-chamber with the main combustion chamber. Depending on the characteristics of the hot jet issued from the pre-chamber, either a *flame jet* (if the pre-chamber flame survives heat loss and high stretch when passing the orifice) or a *hot jet* containing the combustion products only (if the pre-chamber flame extinguishes when passing the orifice) can ignite the main chamber [11–14]. These two fundamentally different ignition mechanisms motivated the authors to investigate how flame propagates through microchannels of various geometries. The nozzles used in practical natural gas engines have diameters ranging from 0.5 to 2 mm. Nevertheless, the term "microscale combustion" is often used when the characteristic length scale is on the same order as the "quenching distance" which is typically a few millimeters [5, 8]. Since most of the channel diameters are within 1–4 mm in the present study, we have used the term "microchannel" throughout this study.

Previous experimental, computational, and theoretical work have revealed rich physics of micro- and mesoscale flame propagation. Many interesting phenomena have been observed, such as flame bifurcation, dynamic oscillating flame, and spinning flames [15–17]. Nevertheless, nearly all previous studies on flame propagation in narrow channels were focused on straight channels or slightly curved channels. Very few have examined converging-diverging (C-D) channels [18]. In a recent study by Biswas and Qiao [14], it was found that using supersonic hot jets generated by using C-D nozzles can ignite leaner mixtures in the pre-chamber hot-jet ignition system, leading to ultra-low emissions and higher combustion efficiency. Thus, studying the flame dynamics through C-D microchannels is of great importance for the understanding of mechanisms of turbulent hot-jet ignition by using C-D nozzles. The main purpose of this study was to conduct this study of flame dynamics through C-D microchannels. Straight microchannels will also be used for comparative studies.

An experiment was developed to study CH_4/air premixed flames passing through straight and C-D microchannels. The primary goal was to find out whether the flame can survive or extinguish while passing through the channels. The influences of the equivalence ratio and channel geometry were studied. Dynamic behavior of flame propagation inside the channels was determined using direct imaging and high-speed CH* chemiluminescence, where flame shape, propagation speed, cyclic oscillatory motions, and local extinction behavior were observed. Numerical simulations of the flames passing through the microchannels were also performed, and the results were used to explain the experimental observations and to help identify flame extinction mechanisms.

8.2 Experimental Methods and Numerical Models

8.2.1 Experimental Methods

The schematic of the experimental setup is shown in Fig. 8.1a. Two identical cubic-shaped carbon steel chambers ($3'' \times 3'' \times 3''$) were connected by a transparent cylindrical quartz micro-tube (type GE124, 86% UV transparent) of 10 cm length. The internal diameter (ID) of the quartz tubes/channels was varied from 1 to 10 mm. The channel dimensions (channel diameter, d; channel length, L; and throat diameter, d_t for C-D channels) are listed in Table 8.1. For each ID, both a straight channel and a C-D channel were tested. For the C-D channels, two different exit-to-throat area ratios (AR) of 4 and 9 were used.

The combustion chamber on the left-hand side of Fig. 8.1b, which mimics a pre-chamber in gas engines with a turbulent jet ignition system, was used to generate a stably propagating laminar flame entering the microchannel. The CH_4/air mixture in the combustion chamber was ignited by using an electric spark created by a 0–40 kV capacitor discharge ignition (CDI) system. The inlet of the microchannels had a bell-shaped opening to accommodate smooth entrance of the flame. The cubic-shaped chamber on the right side was the settling chamber, which was filled with inert gasses and served to release pressure after combustion. A ball valve located at the entrance of the settling chamber was closed during the experiment. Thus, the combustion chamber and the microchannel essentially formed a constant volume system.

High-speed CH^* chemiluminescence, high-speed direct luminosity imaging, and infrared imaging were used to visualize flame propagation in the combustion chamber and the microchannel. A high-speed camera (Vision Research Phantom v7.1), along with video-scope gated image intensifier (VS4-1845HS) with 105 mm UV lens, was utilized to detect CH^* signals at a narrow band 431 ± 12 nm detection limit. The intensifier was externally synced with the camera via a high-speed relay and acquired images at the same frame rate (up to 16,000 fps) with the Phantom camera. The pressure of the combustion chamber and the pressure at the microchannel exit were recorded using high-resolution ($\sim$10 kHz) Kulite (XTEL-190) pressure transducers combined with NI-9237 signal conditioning and pressure acquisition module via the LabVIEW software.

8.2.2 Numerical Models

To provide a quantitative understanding of the dynamics and extinction mechanisms of flames passing through the microchannels, we numerically simulated the transient combustion process inside the combustion chamber and the microchannel. To reduce the computational cost, 2D axisymmetric simulations were performed by transforming the cubic combustion chamber into a cylinder with the same volume.

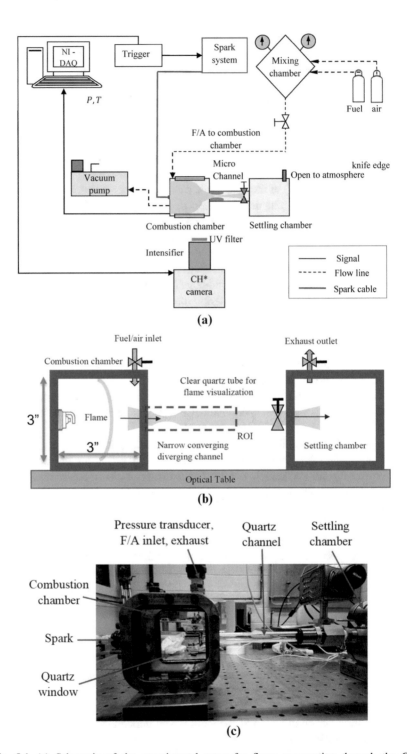

Fig. 8.1 (a) Schematic of the experimental setup for flame propagation through the C-D microchannel, (b) the combustion chamber and the settling chamber, (c) image of the combustion chamber and microchannel assembly

Table 8.1 Dimensions of the microchannels

Channel #	d (mm)	L (cm)	d_t (mm)	Area ratio
1	1	10	0.5	4, 9
2	2	10	1	4, 9
3	3	10	1.5	4, 9
4	4	10	2	4, 9
5	6	10	2	4, 9
6	10	10	5	4, 9

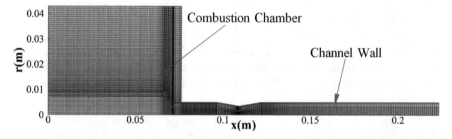

Fig. 8.2 The computational domain and the mesh for the microchannel flames with conjugate heat transfer

The effect of this simplification is expected to be small on the focused study of the flame dynamics inside the microchannels. The simulation results confirmed that the simplified simulations were able to capture the flame propagation inside the combustion chamber reasonably far away from the combustion chamber walls (see Figs. 8.3 and 8.4). The employed computational domain and mesh are shown in Fig. 8.2. Constant volume laminar combustion was assumed. The unstructured computational mesh contained about 100,000 nodes. To examine the effect of heat loss on flame propagation, two thermal boundary conditions were applied to the walls of the microchannels: adiabatic and conjugate heat transfer (fused quartz properties: density 2200 kg/m^3, thermal conductivity 1.38 W/m·K, and specific heat 740 J/kg·K). The convection heat transfer coefficient on the outer surface of the channel wall was estimated to be 10 W/m^2-K based on natural convection heat transfer surrounding a cylinder [19], for imposing the thermal boundary condition for the conjugate heat transfer analysis. At the beginning of the simulations, an initial hot spot of 3 mm in diameter at location $(x, r) = (3$ mm, 0 mm) was provided to initiate flame. The ANSYS Fluent 17.1 with the pressure-based solver was used to conduct the simulations. The chemical mechanism employed is the detailed GRI-Mech 1.2 [20] with 32 species and 175 reactions. In situ adaptive tabulation (ISAT) [21] with an error tolerance 10^{-4} was used to accelerate the numerical integration of chemical reactions. Second-order schemes were used for the temporal and spatial discretization in the simulations.

8.3   Results and Discussion

8.3.1   *Flame Propagation in the Combustion Chamber*

We first examined the flame behavior in the combustion chamber before it entered the microchannel. The CH* chemiluminescence images of flame propagation in the combustion chamber are shown in Fig. 8.3 (top), and the simulated flame temperature contours are shown on the bottom of Fig. 8.3. As the spark, which was located on the center-left side of the chamber, ignited the premixed CH_4/air, a hemispherical-shaped laminar flame was observed to propagate outwardly. Since CH* is generated on the flame front where heat release is maximum, the strongest signals indicate the location of the flame front. As the flame approached near the entrance of the microchannel, the flame tip converted into an elongated-shaped flame front. The simulation captured the flame front indicated by the temperature contour

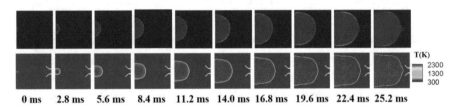

0 ms   2.8 ms   5.6 ms   8.4 ms   11.2 ms   14.0 ms   16.8 ms   19.6 ms   22.4 ms 25.2 ms

Fig. 8.3 High-speed CH* chemiluminescence images of flame propagation (top) and numerically simulated temperature contour (bottom) in the combustion chamber

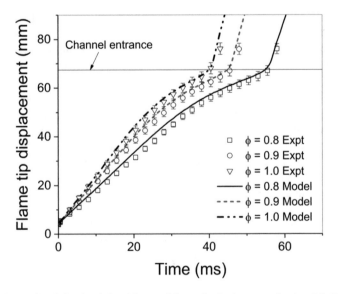

Fig. 8.4 Measured and simulated time history of flame tip displacement for $\phi = 0.8$, 0.9, and 1.0

reasonably well. Slight difference in the predicted flame front curvature from the measurement is expected to be caused by the simplified 2D simulations discussed in Sect. 8.2.2.

Figure 8.4 compares the time history of the measured and simulated flame tip displacement for three equivalence ratios. After the initial transient of spark ignition, the flame propagated nearly at a constant speed (laminar flame speed) inside the chamber, as indicated by the straight lines in the displacement versus time diagram for time less than 30 ms. The slope of the displacement versus time diagram is higher for $\phi = 1.0$ than $\phi = 0.9$ and 0.8. This is expected since the laminar flame speed is maximum near stoichiometric condition. As the flame entered the microchannel, however, the flame tip velocity increased rapidly indicated by the curves above the "channel entrance" line. This acceleration was caused by high-pressure combustion products pushing the unburned fuel/air mixture through the smaller diameter channel. The simulations accurately reproduced the measurements of the flame tip displacement inside the combustion chamber, which confirms the minor effect of using simplified 2D simulations for the combustion chamber.

8.3.2 Flame Dynamics in Microchannels

Once the flame enters the microchannel, depending on the channel geometry and the equivalence ratio, the flame dynamics vary. Figure 8.5a–f compares the time sequence of CH* chemiluminescence images of flame propagation process through straight and C-D channels with ID = 10 mm, 2 mm, and 1 mm, respectively.

Figure 8.5a shows flame propagation through a 10 mm straight channel. The finger-shaped flame front steadily passed through the channel. However, in the C-D channel of the same inlet diameter and with AR = 4, as seen from Fig. 8.5b, the flame became unstable after passing through the throat. The hemispherical shape of the flame front vanished, and the flame became turbulent as evident from the CH* emission. As a result of turbulent flame propagation, the time to travel through the entire channel length was shorter for the C-D channel (2.3 ms) compared to the straight channel (4.2 ms).

Flame behavior changed with a decrease in channel diameter. Figure 8.5c and d present flame propagation under the stoichiometric condition ($\phi = 1$) through a 2 mm microchannel – both straight and C-D. The flames became weaker, as evident from the weaker CH* signals and the shorter length of the reacted (burned) zone, likely due to both increased heat loss to the channel walls and increased flame stretch. The flame remained the finger-liked shape while passing through the straight channel. After passing through the C-D section, however, the flame width shrunk, and it could not occupy the entire channel width. This could be attributed to increasing stretch due to higher velocity at the throat. Unlike the 10 mm case, the flame remained laminar after passing the throat. For straight channels, the propagation speed (location of flame front as a function of time) was nearly constant.

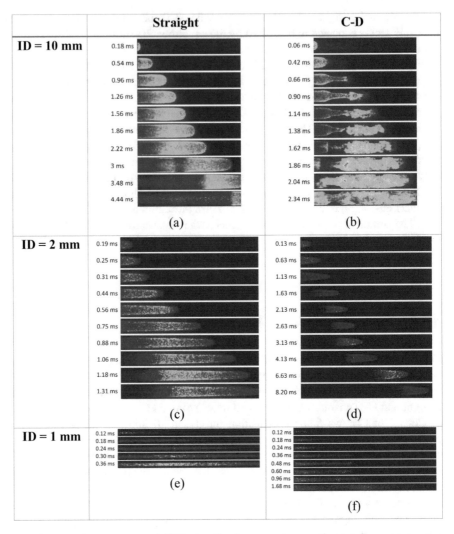

Fig. 8.5 The time sequence of CH* chemiluminescence images showing flame propagation process through different diameters of straight and C-D channels with AR $= 4$, $\phi = 1$

Nevertheless, for C-D channels, the flame front location was not linear with time, especially after passing through the throat the flame slowed down.

The time sequence of flame propagation through a 1 mm microchannel is shown in Fig. 8.5e and f. A thin, sharp-tip flame passed through the straight channel. Depending on the equivalence ratio, the flame barely passed and extinguished in most cases inside the 1 mm microchannel. The flame could not pass when $\phi < 0.9$ or $\phi > 1.2$.

Unlike the 10 mm channel, for the 2 mm (as well as the 1 mm channel), the time taken for the flame to travel through the channel length increased significantly for the C-D channel (8.2 ms) compared to the straight channel (1.3 ms). Due to the presence of the C-D section, it was expected for the flame to pass through the channel faster. This holds true for higher channel diameter (10 mm). Passing through the C-D section of the 10 mm channel, the flame became turbulent and propagated faster. For 2 mm or 1 mm channel diameter, flame front was stretched after passing through the throat. Higher stretch slowed down the flame after it passed the throat for the 1 and 2 mm C-D microchannels.

8.3.3   Extinction and Reignition Phenomena

Flame dynamics becomes interesting as we move away from the stoichiometric condition for smaller channels. Figure 8.6 compares the flame propagation pattern through two C-D channels with the same aspect ratio but different diameters. The equivalence ratio for the two cases was the same. For the 3 mm C-D channel (a), flame was extinguished after passing through the throat, likely due to heat losses and high stretch. However, for the 4 mm C-D channel (b), the flame extinguished partially just at the downstream of the throat. It was reignited again once it arrived at the straight section in the channel.

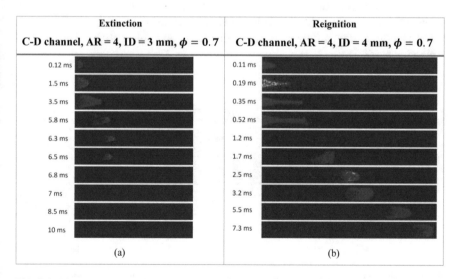

Fig. 8.6 The time sequence of CH* chemiluminescence images showing flame extinction and reignition processes

8.3.4   Three Patterns: Passing, Partially Passing, and Not Passing

The flame passing behavior in the microchannels depends on the channel geometry and the equivalence ratio of CH_4/air mixtures and can be divided into three categories:

Passing: When a flame runs through the entire channel stably without any oscillations or retardation inside the channel.

Partially passing: When a flame passes the channel after retardation or oscillations. The flame seems to stand still, oscillate, or extinguish momentarily inside the channel and then reignites and passes through the channel eventually.

Not passing: When the flame extinguishes in the channel and is unable to pass the entire length.

Figure 8.7 summarizes the three flame passing behaviors as functions of channel diameter and equivalence ratio for straight and C-D channels with AR = 4 and 9, respectively. Overall, for both straight and C-D channels, when the equivalence ratio and channel diameter decrease, the flame pattern shifted from passing to not passing. Additionally, a flame could pass through a straight channel more easily than a C-D channel for a given diameter and equivalence ratio. Moreover, as AR was increased from 4 to 9, the region for passing became even narrower, indicating the flame was more likely to extinguish at higher values of AR. This was because as the throat diameter decreased, the flame was exposed to higher stretch, leading to extinction more easily. The simulations with conjugate heat transfer confirmed the identification of the different passing behaviors, and this will be discussed in following sections.

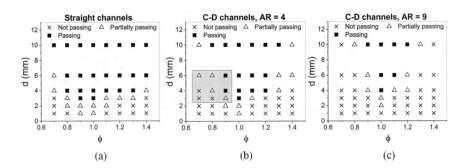

Fig. 8.7 Diagram of three different flame behaviors as functions of channel diameter and equivalence ratio for (**a**) straight channels, (**b**) C-D channels with AR = 4, (**c**) C-D channels with AR = 9, based on the experimental observations

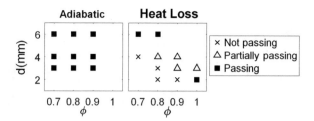

Fig. 8.8 Predicted flame behavior with and without heat loss

8.3.5   Effect of Heat Loss

As discussed earlier, heat loss and high stretch are the two mechanisms likely responsible for flame extinction in the straight and C-D microchannels. Fundamentally, when microchannel diameter decreased, both the effects of stretch and heat loss increased to lead to potential flame extinction. In this section, we numerically examined the effect of heat loss on the flame propagation process to isolate its effect from the stretch effect during extinction. The test conditions marked by a gray-shaded box in Fig. 8.7b were selected for simulations corresponding to a C-D channel with AR $= 4$. Two sets of simulations were conducted, one with adiabatic wall boundary condition and the other considering heat loss through the channel wall using conjugate heat transfer as discussed in Sect. 8.3. Figure 8.8 compares the flame pattern diagram obtained from the simulations with and without heat loss. From the experimental Fig. 8.7b, it was observed that the critical equivalence ratio moved toward $\phi = 1.0$ when the channel diameter decreased from 6 to 2 mm. Numerical modeling with conjugate heat transfer was able to predict this similar trend, while the adiabatic case predicted passing for all conditions. In other words, pure stretch effect (without heat loss effect) hardly affects the flame passing behaviors in Fig. 8.8. This implies that heat loss was a dominant factor to cause the flame to either partially pass or not pass through the C-D microchannels.

To further illustrate the significance of heat loss, Fig. 8.9 shows the time sequence of predicted OH contours in the microchannel from two simulations for a C-D channel with ID $= 3$ mm, AR $= 4$, and $\phi = 0.8$, with adiabatic boundary condition and with the conjugate heat transfer. For the adiabatic case, the flame passed through the channel smoothly, and the mass fraction of OH was high behind the flame front. For the conjugate heat transfer case, the flame entering the channel seemed to be separated from the combustion chamber combustion products as seen from the OH contours. The annihilated OH behind the flame front in the conjugate heat transfer case was probably caused by the significant heat loss to the wall. After the throat, the flame front fluctuated for some time and eventually extinguished downstream for the conjugate heat transfer case, due to lack of weakened thermal support behind the flame front.

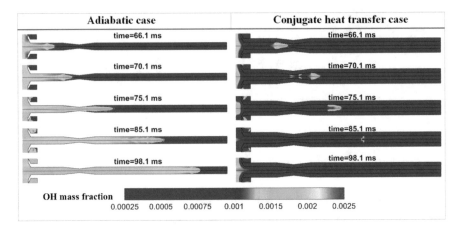

Fig. 8.9 The time sequence of the computed OH contours showing the effect of heat loss (C-D, ID = 3 mm, AR = 4, and $\phi = 0.8$)

8.3.6 Effect of Stretch

From Sect. 8.3.5, we were able to identify the significant effect of heat loss on flame extinction when the channel diameter was reduced. The effect of flame stretch seemed to be the secondary effect on flame extinction for study cases. In this section, we isolated the effect of stretch on the flame propagation pattern from the effect of heat loss. We compared three cases of C-D microchannels of two different area ratios, and the other one was a straight microchannel. All the conditions such as ID and ϕ were kept identical, so that the heat loss effect can be approximately fixed to isolate the effect of stretch. We expected that the major difference between the two cases was the stretch rate which was higher in the C-D channel than in the straight channel. From the experimental results shown in Fig. 8.7, it was observed that for the same channel ID, the critical equivalence ratio in the C-D microchannel was closer to the stoichiometric condition compared to the straight channel. This observation demonstrated the effect of stretch on flame propagation behaviors. Flames subjected to a higher stretch rate were more vulnerable to extinction. Similar effect of stretch was observed from the numerical simulations (results not shown). Figure 8.10 compares experimentally measured flame tip velocity magnitude for two C-D channels with area ratio of 4 and 9 and a straight channel with the same internal diameter, ID = 3 mm and equivalence ratio, $\phi = 0.8$. Since the stretch rate is proportional to the velocity gradient, the flame tip velocity provided a qualitative measure of the effect of stretch. The flame tip velocity oscillations in the C-D channels were much higher compared to the straight channel. Among two different area ratios, AR = 9 had higher velocity oscillations. Hence, the flame in C-D channel of AR = 9 was more susceptible to higher stretch rate. Oscillating flame in microchannel had been observed by several researchers [7, 8, 22, 23]. Flame oscillation behavior in microchannel could be attributed to the competition between wall

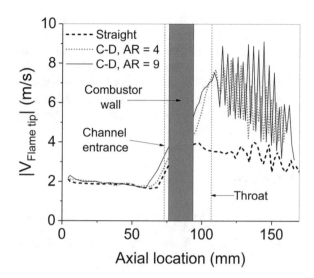

Fig. 8.10 Flame tip velocity in C-D channels and straight channel for ID = 2 mm and $\phi = 0.8$

heat loss and heat gain from upstream combustion. Flame shrunk after losing heat through the wall. As the flame shrunk, the effect of heat loss diminished, and the flame could be stabilized by enhanced burning rate. This periodic behavior continued until the flame could not withstand the stretch or if it experienced excessive heat loss.

From Sects. 8.3.5 and 8.3.6, we were able to isolate the effect of stretch from heat losses to characterize flame extinction behavior inside the microchannels. Comparing the effects of heat loss and stretch in the microchannel flames, heat loss had a more noteworthy influence on flame extinction, as discussed in Sect. 8.3.5.

8.4   Conclusions

This study describes the dynamics of premixed CH_4/air flame propagation through straight and C-D microchannels. The key findings are summarized below:

1. While a finger-shaped flame passed through the straight channel, the shape of the flame passing through the C-D channels changed with the diameter. For a higher channel diameter, the flame became turbulent passing through the divergent section. For a smaller diameter channel, the flame remained laminar and was likely to extinguish near the throat and was reignited at the downstream.
2. Three flame behaviors were observed – passing, partially passing, and extinguish, depending on the channel geometry and equivalence ratio. In general, flames were easier to extinguish in C-D channels than in straight channels for a fixed channel diameter and equivalence ratio. Additionally, flames were harder to survive in C-D channels with larger exit-to-throat area ratio (higher contraction).

3. Both heat loss and flame stretch were factors that could cause flame extinction in small C-D channels. Simulations with and without heat loss reveal that heat loss was the primary reason for the partial passing of flames through the microchannel. The isolated stretch effect was also studied with the heat loss effect fixed. Due to the stretch effect, the critical equivalence ratio in the C-D microchannel was closer to the stoichiometric condition than the straight channel. Lastly, flame oscillations were observed in both experiment and simulations for smaller microchannels.

References

1. Fernandez-Pello, A.C.: Micropower generation using combustion: issues and approaches. Proc. Combust. Inst. **29**, 883–899 (2002)
2. Yetter, R., Yang, V., Aksay, I., Dryer, F.: Meso and Micro Scale Propulsion Concepts for Small Spacecraft. STAR. **44**(23), 1–51 (2006)
3. Dunn-Rankin, D., Leal, E.M., Walther, D.C.: Personal power systems. Prog. Energy Combust. Sci. **31**(5–6), 422–465 (2005)
4. Abas, N., Kalair, A., Khan, N.: Review of fossil fuels and future energy technologies. Futures. **69**, 31–49 (2015)
5. Ju, Y., Maruta, K.: Microscale combustion: technology development and fundamental research. Prog. Energy Combust. Sci. **37**(6), 669–715 (2011)
6. Ju, Y., Xu, B.: Theoretical and experimental studies on mesoscale flame propagation and extinction. Proc. Combust. Inst. **30**(2), 2445–2453 (2005)
7. Nakamura, H., et al.: Bifurcations and negative propagation speeds of methane/air premixed flames with repetitive extinction and ignition in a heated microchannel. Combust. Flame. **159** (4), 1631–1643 (2012)
8. Maruta, K.: Micro and mesoscale combustion. Proc. Combust. Inst. **33**(1), 125–150 (2011)
9. Maruta, K., et al.: Characteristics of combustion in a narrow channel with a temperature gradient. Proc. Combust. Inst. **30**(2), 2429–2436 (2005)
10. Kim, N.I., et al.: Flammability limits of stationary flames in tubes at low pressure. Combust. Flame. **141**(1–2), 78–88 (2005)
11. Biswas, S., Qiao, L.: A numerical investigation of ignition of ultra-lean premixed H2/air mixtures by pre-chamber supersonic hot jet. SAE Int. J. Engines. **10**(5), 2231–2247 (2017)
12. Biswas, S., Qiao, L.: Ignition of ultra-lean premixed H2/air using multiple hot turbulent jets generated by pre-chamber combustion. Appl. Therm. Eng. **106**(5), 925–937 (2017)
13. Biswas, S., et al.: On ignition mechanisms of premixed CH4/air and H2/air using a hot turbulent jet generated by pre-chamber combustion. Appl. Therm. Eng. **106**, 925–937 (2016)
14. Biswas, S., Qiao, L.: Prechamber hot jet ignition of ultra-lean H2/air mixtures: effect of supersonic jets and combustion instability. SAE Int. J. Engines. **9**(3), 1584–1592 (2016)
15. Alipoor, A., et al.: Asymmetric hydrogen flame in a heated micro-channel: role of Darrieus–Landau and thermal-diffusive instabilities. Int. J. Hydrog. Energy. **41**(44), 20407–20417 (2016)
16. Wang, Y., et al.: The impact of preheating on stability limits of premixed hydrogen–air combustion in a microcombustor. Heat Transfer Eng. **33**(7), 661–668 (2012)
17. Sánchez-Sanz, M., Fernández-Galisteo, D., Kurdyumov, V.N.: Effect of the equivalence ratio, Damköhler number, Lewis number and heat release on the stability of laminar premixed flames in microchannels. Combust. Flame. **161**(5), 1282–1293 (2014)
18. Yang, W., et al.: Experimental and numerical investigations of hydrogen–air premixed combustion in a converging–diverging micro tube. Int. J. Hydrog. Energy. **39**(7), 3469–3476 (2014)
19. Churchill, S.W., Chu, H.H.S.: Correlating equations for laminar and turbulent free convection from a horizontal cylinder. Int. J. Heat Mass Transf. **18**, 1049–1053 (1975)

20. GRI-Mech 1.2. http://www.me.berkeley.edu/gri_mech/
21. Pope, S.B.: Computationally efficient implementation of combustion chemistry using in situ adaptive tabulation. Combustion Theory Model. **1**(1), 41–63 (1997)
22. Mauger, C., et al.: Velocity measurements based on shadowgraph-like image correlations in a cavitating micro-channel flow. Int. J. Multiphase Flow. **58**, 301–312 (2014)
23. Fan, A., et al.: Experimental investigation of flame pattern transitions in a heated radial micro-channel. Appl. Therm. Eng. **47**, 111–118 (2012)

Chapter 9
Summary and Outlook

Contents

9.1 Summary

The ignition characteristics of methane/air and hydrogen/air mixtures using a hot turbulent jet generated by pre-chamber combustion were studied from a fundamental point of view. The existence of two different ignition mechanisms, namely, jet ignition and flame ignition, was found. Jet ignition produced a jet of hot combustion products (which means the pre-chamber flame is quenched when passing through the orifice); flame ignition produced a jet of wrinkled turbulent flames (the composition of the jet is incomplete combustion products containing flames). As the orifice diameter increased, the ignition mechanism switched to flame ignition, from jet ignition. With the increase in pressure, flame ignition became more prevalent. The ignition took place on the side surface of the hot jet during the jet deceleration process for both mixtures. A critical global Damköhler number, Da_{crit}, defined as the limiting parameter that separated ignition from no ignition, was found to be 140 for methane/air and 40 for hydrogen/air. All possible ignition outcomes were compared on the turbulent combustion regime diagram. Nearly all no-ignition cases fell into the broken reaction zone, and jet and flame ignition cases mostly fell within the thin reaction zones.

Current research demonstrated a novel, inexpensive, easy to setup two-camera SIV technique that could resolve exceptionally high flow velocities. Statistical assessment of SIV techniques was performed for a high-velocity helium jet at two different Reynolds numbers, $Re_d = 11,000$ and $Re_d = 22,000$. The velocity field

© Springer International Publishing AG, part of Springer Nature 2018   197
S. Biswas, *Physics of Turbulent Jet Ignition*, Springer Theses,
https://doi.org/10.1007/978-3-319-76243-2_9

obtained by horizontal knife-edge schlieren with 40% cutoff and shadowgraph agreed well with the PIV results. Vertical knife-edge schlieren with 40% cutoff performed poorly due to inconsistent signal content. Optimization of SIV processing parameters, diverse types of image filtering, image restoration, and noise reduction techniques useful for SIV techniques were discussed in detail. The present study which consists of an open-source PIV processing algorithm with a novel two-camera, easy-to-set up SIV technique with a detailed description of image preprocessing, flow field post-processing, and their statistical assessment, therefore, presented a solution to resolve velocity fields of a wide range of turbulent flows.

A vital finding of this research was the extension of lean limit and lower ignition delay of the ultra-lean hydrogen/air mixture by using a supersonic jet. Results show ignition initiates from the side surface of the hot jet. Due to the presence of shock structures, supersonic jet exit temperature was higher than a subsonic jet. The increase in the static temperature behind the shocks thus escalated ignition probability and reduced the lean limit. The mechanism why the supersonic jets could extend the lean flammability limit of hydrogen/air mixture was explained using numerical modeling. Due to higher velocity and vorticity, the supersonic jets could mix with the cold unburned hydrogen/air more efficiently than subsonic nozzles. Simultaneously, the static temperature of the supersonic jets increased after each shock, and after the final strong shock, the temperature rise was significant. Main chamber ignition was initiated from this high-temperature region. These two phenomena together raised the possibility of ultra-lean ignition using supersonic jets over subsonic jets. This finding could help us better control the ignition location and ignition delays and design a better pre-chamber for lean combustion.

Thermoacoustic combustion instability arises at the ultra-lean condition. Thermoacoustic instability of ultra-lean premixed hydrogen/air was characterized using experiment and modeling. The strongest mode always corresponded to the first longitudinal (1 L) mode of the system. The first transverse (1 T) mode was chamber geometry dependent. The higher-order modes were the complex coupled mode of the combustor. Unstable modes from 3D LEE matched better with the experimental data as compared to 1D LEE. The first unstable mode of combustion instability always corresponded to the longitudinal mode of the system. Transverse and complex mixed modes arose at lower equivalence ratios. A supercritical bifurcation occurred, and instability triggered in for $\phi < 0.5$. The frequency of the first longitudinal (1 L) mode decreased with decreasing equivalence ratio due to change in adiabatic flame temperature. The frequency of the 1 L mode ranged from 1400 to 2200 Hz. Strain rates fluctuated along the oscillating flame edge. This caused higher strain at antinodes and lower strain at nodes of pressure perturbation cycle. The maximum strain rate reached four- to fivefold of the minimum strain rate in a typical pressure perturbation cycle at the lean flammability limit for hydrogen/air. At the lean flammability limit, the fluctuating pressure reached 25% of the mean pressure. Dynamic mode decomposition (DMD) analysis was performed on OH* chemiluminescence images to identify the physical significance of the dominant unstable modes in the combustor. The 1 L mode always corresponded to the heat release mode of the combustor.

Ignition characteristics of ultra-lean premixed hydrogen/air by multiple hot turbulent jets in a dual combustion chamber system were studied. Compared to single jet, multiple jets (straight or angled) did not extend the lean flammability limit of H_2/air, given both systems have the same total nozzle area. However, the ignition probability improved significantly near the lean flammability limit by using multiple jets. This is due to the cumulative ignition probability which is higher for multiple hot-jet ignitions compared to single jets. For multiple jets, the spark location and the fuel/air equivalence ratio inside the pre-chamber had a deterministic effect on the ignition pattern in the main chamber. If the pre-chamber spark was located such that the effective length to diameter ratio exceeded the critical L/D ratio of 2, the side jets ignited the main chamber first. Otherwise, the middle jet or all the jets ignited the main chamber depending on the pre-chamber equivalence ratio. For a specific spark location, a richer hydrogen/air mixture tended to increase main chamber ignition by all jets. Overall, the effect of pre-chamber equivalence ratio and the spark location was strongly coupled. The numerical simulation results showed that the flame shape inside the pre-chamber when approaching the nozzles determined the ignition pattern, e.g., ignition was started by the side jets or the middle jet. For pre-chambers having an effective L/D over 2, flame front inverted and became a tulip-shaped flame, promoting ignition by the side jets. Both experiments and simulations showed that if the spark location was too close to the nozzle entrance, ignition was initiated by laminar or slightly turbulent flame jets. The main chamber burn rate increased with multi-jets compared to a single jet. Moreover, angled multi-jets increased the burn rate even more compared to straight multi-jets by enhancing turbulence and mixing inside the main combustion chamber.

The effect of jet impingement on ultra-lean ignition physics was studied in detail. Impinging jet ignition could extend the flammability limit of the lean H_2/air mixture in the main chamber, as compared to typical jet ignition. This was because impinging jet ignition created a high-temperature, stagnation zone near the impinging point and ignition started from this zone. Enhanced mixing due to increase in vorticity near the impinging region allowed surrounding unburned fuel/air to penetrate and enter into the stagnation region. A high dynamic temperature grew in the stagnation region facilitated ignition. This was the reason why impinging jet could extend the lean flammability limit of H_2/air. Furthermore, impinging jet ignition decreased the ignition delay in the main chamber. The dominating factor in the impinging jet ignition was the impinging distance and impinging angle, which determined the probability of impinging jet ignition. A higher impinging distance did not result in impinging jet ignition. Rather, ignition occurred by typical turbulent jet ignition mechanism. With the increase in the impinging angle, the stagnation point shifted away from the jet centerline, and thus the probability of impinging jet ignition decreased.

Dynamics of premixed methane/air flame propagation through straight and C-D microchannels were investigated in depth. Three flame behaviors were observed – passing, partially passing, and extinguish, depending on the channel geometry and equivalence ratio. In general, flames were easier to extinguish in C-D channels than in straight channels for a fixed channel diameter and equivalence ratio. Additionally, flames were harder to survive in C-D channels with larger exit-to-throat area ratio (higher contraction). Both heat loss and flame stretch were factors that could cause flame extinction in small C-D channels. Simulations with and without heat loss reveal that heat loss was the primary reason for the partial passing of flames through the microchannel. The isolated stretch effect was also studied with the heat loss effect fixed. Due to the stretch effect, the critical equivalence ratio in the C-D microchannel was closer to the stoichiometric condition than the straight channel. Lastly, flame oscillations were observed in both experiment and simulations for smaller microchannels.

9.2 Outlook for Future

The promise and potential of ultra-lean ignition by a hot turbulent jet generated by pre-chamber combustion are vast. Ultra-lean engine operation has the potential to reduce greenhouse emissions and increase fuel economy and thus needs to be investigated in detail to understand this subject fully. Interesting findings from our fundamental study on hot turbulent jet ignition show the enormous potential of this technology. However, the present research was just a start toward understanding the complex ignition physics in detail. Following are few prominent future research directions that can be pursued.

9.2.1 High-Pressure Jet Ignition

For a heavy-duty engine, the pressure and temperature of the combustion chamber can go well above 50 atmospheric, 800 K. At this pressure and temperature, flame speed, flame thickness, and ignition strain rate also change. In an engine, relevant conditions, at high pressure and high temperature, the flame dynamics become very different. The characteristics of the hot turbulent jet will change at engine conditions. Also, it would be interesting to see the effect of thermal stratification on jet ignition inside the engine. Even though the characteristic non-dimensional numbers will be unchanged, the ignition physics may be governed by some other non-dimensional numbers such as Karlovitz number, Lewis number, and Markstein number at such high pressure. It would be interesting to study the turbulent jet ignition at high pressure and high temperature.

9.2.2 High-Speed Laser Diagnostics

Initiation of ignition is a complex phenomenon. There exists a thermal runway during which radicals initiation happens. If the radical generation process sustains through chain-branching reactions, it will lead to ignition. However, this entire phenomenon happens in the order of few hundred milliseconds. To resolve such ultrafast phenomena, we need high-speed diagnostics. To measure the presence of fuel stratification, Rayleigh scattering can be used. To better understand the shape and location of the ignition kernels, high-speed OH and CH laser-induced fluorescence (LIF) can be used. The jet temperature is a key parameter in the turbulent jet ignition. High-speed femtosecond coherent anti-Stokes Raman spectroscopy (CARS) can be used to measure jet temperature. A high-speed volumetric particle image velocimetry (PIV) would also be helpful to understand the flow field during ignition.

9.2.3 Effect of Different Pre-chamber Fuels and Fuel Stratification

In our current research, we have studied the hot turbulent jet ignition characteristics of methane and hydrogen. It would be interesting to see the effect of different fuels on jet ignition characteristics. A mixture of natural gas and hydrogen will also be interesting to study. Before the hot combustion products enter the main chamber as a hot jet, it pushes unburned stoichiometric fuel/air mixture into the ultra-lean main chamber. This creates a fuel stratification around the hot jet. This unburned stoichiometric or near-stoichiometric fuel/air mixture created localized fuel-enhanced pockets. The ignition probability would be higher in these pockets. If the fuel diffusion time is higher than the ignition time, it will be interesting to explore the ignition characteristics starting from these locally fuel-stratified regions.

9.2.4 Uncertainty Analysis of Schlieren Image Velocimetry

Schlieren image velocimetry (SIV) is a novel seedless velocimetry measuring technique that can be applied to many types of turbulent flows. The underlying physics of SIV is very different from particle image velocimetry (PIV). In SIV techniques, the eddies in a turbulent flow field serve as PIV "particles." These "particles" can then be cross-correlated to find the velocity field. Since SIV is a quantitative measurement, the inherent uncertainties associated with the SIV data must be objectively assessed. The uncertainty is SIV measurement arises in two stages. The first set of uncertainties comes from the optical setup of schlieren/ shadowgraph. The mirror, lens, knife edge, and the quality of collimated light beam affect the final schlieren image produced. Thus, we need to assess the effect of individual optical elements on schlieren imaging. One way to do it is by Fourier optics. Using Fourier optics, we can write the optical transfer function of each component and construct a mathematical framework to evaluate the errors arising from the optical setup. The second set of uncertainties arises at the processing stage. The eddy sizes change within the flow field, which means the particle sizes for SIV calculations also change. This can induce errors in the cross-correlation statistics. The errors arising from the processing of SIV needs to be addressed as well.

Appendices

Appendix A: Partial Pressure Calculation

According to Dalton's law of partial pressure, the total pressure, P_{total}, exerted by a gas mixture composed of N different gases, is equal to the sum of the partial pressure of individual gas, p.

$$P_{total} = \sum_{i=N} p_i \qquad (A.1)$$

For an ideal gas mixture, the ratio of a gas component's partial pressure to total pressure is equal to the mole fraction of the component, X_i.

$$X_i = \frac{p_i}{P_{total}} \qquad (A.2)$$

This partial pressure technique was used to control the fuel/air ratio in the constant volume combustion chamber. Consider preparing mixtures of methane/air and hydrogen/air of equivalence ratio ϕ. The fuel/air mixture compositions can be written as

$$\phi CH_4 + 2(O_2 + 3.76N_2) \rightarrow Products \qquad (A.3)$$
$$\phi H_2 + 0.5(O_2 + 3.76N_2) \rightarrow Products \qquad (A.4)$$

The mole fractions of methane and hydrogen to achieve an equivalence ratio of ϕ can be expressed in terms of their corresponding partial pressures.

$$X_{CH_4} = \frac{\phi}{\phi + 2(1 + 3.76)} = \frac{p_{CH_4}}{P_{total}} \qquad (A.5)$$

© Springer International Publishing AG, part of Springer Nature 2018
S. Biswas, *Physics of Turbulent Jet Ignition*, Springer Theses,
https://doi.org/10.1007/978-3-319-76243-2

$$X_{H_2} = \frac{\phi}{\phi + 0.5(1 + 3.76)} = \frac{P_{H_2}}{P_{total}} \qquad (A.6)$$

For given equivalence ratio, ϕ, and initial pressure, P_{total}, partial pressures of fuel and similarly air were calculated using Eqs. A.5 and A.6. The mixing chamber was filled up by fuel/air using partial pressure method.

Appendix B: Diaphragm Selection

For all jet ignition test conditions, the fuel/air mixture in the pre-chamber was always kept stoichiometric or near stoichiometric. However, main chamber equivalence ratio was varied from $\phi = 0.22 - 0.9$. To separate these two chambers with dissimilar equivalence ratios, a separating diaphragm/membrane was necessary. The diaphragm had to meet the following standards.

1. Sustain a specified pressure difference and then break cleanly when the pressure difference is reached. Since the combustion is dependent on the diaphragm rupture, the rupture time or, in other words, the rupture pressure needs to be extremely repeatable.
2. While the chamber is in a vacuum, the diaphragm should be intact.
3. The diaphragm should be made of a thin material that when it breaks, the broken diaphragm pieces do not affect the flow.
4. When the chamber is heated, the diaphragm should not melt and stick to the chamber.

We bought three different diaphragm sheets with varying thickness from McMaster as listed in Table B.1. The diaphragm rupture times and rupture pressure differences are shown in Fig. B.1 on a typical pre-chamber pressure profile. We did not want the diaphragm to open too soon or too late. Rather we wanted to open as the pre-chamber pressure started moving up and the pre-chamber flame was still away from the nozzle entrance. Thus, we chose diaphragm 2. Diaphragm 2 was a light-weight, 25 ± 1.25-micron thick aluminum sheet (aluminum alloy1100) that we used for all tests.

Table B.1 Different types of diaphragms tested

	Thickness (µm)	Material
Diaphragm 1	13	18-8 stainless steel
Diaphragm 2	25	Aluminum
Diaphragm 3	51	Aluminum

Fig. B.1 Comparison of
rupture time (*t*) and rupture
pressure difference (*dP*) of
three different diaphragms

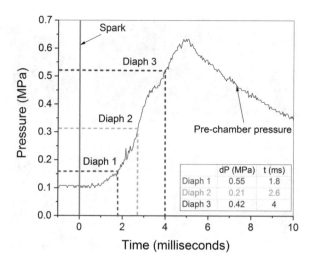

Appendix C: Two-Camera Calibration

In schlieren image velocimetry (SIV), two cameras were used side by side to capture high-speed helium jet. The field of view (FOV) from the two cameras was not the same; rather, FOVs were shifted. Even few pixels of shift can change the final velocity field by drastically. To take care of this issue, a dot pattern and an image were photographed at the location of imaging by two cameras. Then a calibration map was generated based on the difference in FOV. This map was applied to one set of camera images.

An example of such calibration is shown in Fig. C.1. The camera images were subtracted to find the difference in the FOVs. Then a Gaussian mapping was applied to this difference to ease out the abrupt change in mapping function. Then this mapping function was used to rectify the camera 2 images. Then the camera 1 and rectified images of camera 2 were cross-correlated.

Appendix D: Abel Inversion

Abel inversion is necessary to reconstruct 2D velocity field from the line of sight integrated schlieren velocity field. Figure D.1 shows the Abel inversion of a schlieren image of a turbulent jet. Since schlieren retains volumetric information from the flow field, we need to reconstruct the flow field before going into data processing. The inverse Abel transform used in this work has been derived in the following section.

Abel transform of a function $f(r)$

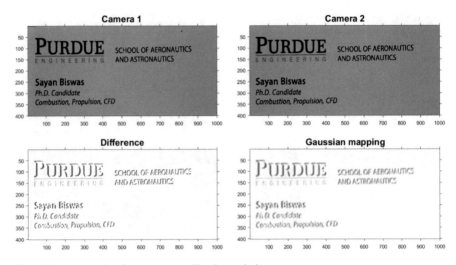

Fig. C.1 An example of two-camera calibration technique

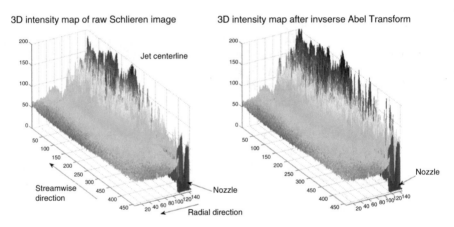

Fig. D.1 Abel inversion of schlieren image of a turbulent reacting jet of Reynolds number $Re_D = 37,200$

$$F(y) = 2 \int_y^\infty \frac{f(r)r\,dr}{\sqrt{r^2 - y^2}} \qquad (D.1)$$

Assume $f(r)$ reaches zero faster than $1/r$, the Abel inversion can be written as

$$f(r) = -\frac{1}{\pi} \int_r^\infty \frac{dF}{dy} \frac{dy}{\sqrt{y^2 - r^2}} \qquad (D.2)$$

We can verify the inverse Abel transform by following method. Assume $u = f(r)$ and $v = \sqrt{r^2 - y^2}$. Then use integration by parts

$$(y) = 2 \int_y^\infty f'(r)\sqrt{r^2 - y^2} \tag{D.3}$$

Differentiating both sides, we get

$$F'(y) = 2y \int_y^\infty \frac{f'(r)}{\sqrt{r^2 - y^2}} dr \tag{D.4}$$

Now put it in the inverse Abel transform Eq. D.1.

$$-\frac{1}{\pi} \int_r^\infty \frac{F'(y)}{\sqrt{y^2 - r^2}} dy = \int_r^\infty \int_y^\infty -\frac{2y}{\sqrt{y^2 - r^2}\sqrt{s^2 - r^2}} f'(s) ds dy \tag{D.5}$$

Using Fubini's theorem which let us compute a double integral using iterated integrals, the last integral can be solved as

$$\int_r^\infty \int_y^\infty \frac{-2y}{\pi\sqrt{y^2 - r^2}\sqrt{s^2 - r^2}} f'(s) ds dy = \int_r^\infty (-1)f'(s) ds = f(r) \tag{D.6}$$

This completes our derivation of inverse Abel transform.

Appendix E: Uncertainty and Error Analysis

Since experimental measurements of a variable contain inaccuracies, it is important to understand these inaccuracies.

E1. Uncertainty for Spectral Radiation Intensity

A derived quantity, R, is a function of measured variables, X.

$$R = f(X_1, X_2, X_3, \ldots, X_i, \ldots, X_N) \tag{E.1}$$

If each variable X has an uncertainty ΔX, the overall uncertainty measuring R can be estimated as

$$\frac{\Delta R}{R} = \sqrt{\sum_{i=1}^{N} \left[\partial \frac{\ln R}{\partial X_i} \Delta X_i\right] 2} \tag{E.2}$$

The spectral radiation intensity, I_λ, was calculated from the blackbody radiation intensity, $I_{b\lambda}$; the flame radiation-induced voltage (detector output voltage), V; and the measured voltage, V_m, using the following equation.

$$I_\lambda = I_{b\lambda}\frac{V}{V_m} \tag{E.3}$$

Thus, the uncertainty in measuring the radiation intensity can be written as

$$\frac{\Delta I_\lambda}{I_\lambda} = \sqrt{\left(\frac{\Delta I_{b\lambda}}{I_{b\lambda}}\right)^2 + \left(\frac{\Delta V}{V}\right)^2 + \left(\frac{\Delta V_m}{V_m}\right)^2} \tag{E.4}$$

We calculated that the uncertainty in the measured voltage, $\Delta V_m/V_m$, is less than 1.5% in experiment. The uncertainty in the flame radiation-induced voltage, $\Delta V/V$, is also less than 7%. The uncertainty in $I_{b\lambda}$ was caused by the uncertainty in the temperature and wavelength calculations during the calibration process and can be determined from the Planck function

$$I_{b\lambda} = \frac{C_1}{\lambda^5\left(exp\left(\frac{C_2}{\lambda T}\right) - 1\right)} \tag{E.5}$$

where $C_1 = 3.742E - 16$ W/m^2 and $C_2 = 14,388$ μmK. The uncertainty in $I_{b\lambda}$ can be estimated as

$$\frac{\Delta I_{b\lambda}}{I_{b\lambda}} = \sqrt{\left(25 - \frac{10C_2}{\lambda T} + \frac{C_2^2}{\lambda^2 T^2}\right)\left(\frac{\Delta\lambda}{\lambda}\right)^2 + \frac{C_2^2}{\lambda^2 T^2}\left(\frac{\Delta T}{T}\right)^2} \tag{E.6}$$

The maximum absolute uncertainty in the blackbody temperature was 0.9 K, and the maximum absolute uncertainty in the wavelength setting was 8 nm. For measured temperature range, 400 K $< T <$ 1600 K, and wavelength range, 1.9 μm $< \lambda <$ 3.6 μm, the total uncertainty for I_λ is less than 6%.

E2. Uncertainty in Jet Exit Velocity

Hot-wire pyrometry technique was performed to measure jet exit temperature. We found that the measured jet exit temperature is within 3.5% of adiabatic flame temperature. Kulite pressure transducer has an uncertainty ±0.1% of its full-scale value. Thus, the relative error for density can be written as

$$\frac{\Delta \rho}{\rho} = \frac{1}{\rho R} \left[\left(\frac{\partial \rho}{\partial P} \right) \sigma_P + \left(\frac{\partial \rho}{\partial T} \right) \sigma_T \right] \tag{E.7}$$

$$\frac{\Delta \rho}{\rho} = \frac{1}{PR} \sigma_P - \frac{P}{T^3} \sigma_T \tag{E.8}$$

This produces an uncertainty of 1.72% in density calculations due to uncertainty in pressure and temperature. Thus, we can safely say that the uncertainty in density calculation due to uncertainty in pressure and temperature is negligible.

At any instance of time, exit velocity can be written as

$$U_0 = CY \sqrt{(P_1 - P_2)/\rho_{\text{mix}}} \tag{E.9}$$

where P_1 and P_2 are pre-chamber and main chamber pressure, respectively. Using functional form f C and Y exit velocity can be written as

$$U_0 = \frac{C_d}{\sqrt{1 - \beta^4}} \left[1 - f(\beta) \left\{ 1 - \left(\frac{P_2}{P_1} \right)^{\frac{1}{\gamma}} \right\} \right] \sqrt{\frac{P_1 - P_2}{\rho_{\text{mix}}}} \tag{E.10}$$

Except for pressure, everything else can be treated as constant, and the exit velocity can be written as

$$U_0 = C_1 \left[1 - C_2 + C_2 \left(\frac{P_2}{P_1} \right)^{C_3} \right] C_4 \sqrt{P_1 - P_2} \tag{E.11}$$

where C_1, C_2, C_3, and C_4 are constants. The standard deviation of the error in velocity can be expressed as

$$\sigma_{U_0}^2 = \left(\frac{\partial U_0}{\partial P_1} \right)^2 \sigma_{P_1}^2 + \left(\frac{\partial U_0}{\partial P_2} \right)^2 \sigma_{P_2}^2 \tag{E.12}$$

$$\sigma_{U_0}^2 = \left[\frac{C_1 C_4 \left[C_2 \left(\frac{P_2}{P_1} \right)^{C_3} - C_2 + 1 \right]}{2 \sqrt{P_1 - P_2}} - \frac{C_1 C_2 C_3 C_4 P_2 \sqrt{P_1 - P_2} \left(\frac{P_2}{P_1} \right)^{C_3 - 1}}{P_1^2} \right] \sigma_{P_1}^2$$

$$+ \left[\frac{C_1 C_2 C_3 C_4 P_2 \sqrt{P_1 - P_2} \left(\frac{P_2}{P_1} \right)^{C_3 - 1}}{P_1} - \frac{C_1 C_4 \left[C_2 \left(\frac{P_2}{P_1} \right)^{C_3} - C_2 + 1 \right]}{2 \sqrt{P_1 - P_2}} \right] \sigma_{P_2}^2$$

$$\tag{E.13}$$

We calculated the relative error σ_{U_0}/U_0 is $\pm 2.8\%$.

E3. PIV Error from Physical Setup

Particle volume fraction dependence of accuracy of PIV results has been discussed below.

$$\text{PIV error} \sim \frac{e}{\sqrt{n}} \tag{E.14}$$

where $n = \frac{C \Delta t A^2}{M^2}$, where n is the number of particles in each interrogation window, C is particle concentration (kg/s-m^3 air flow), Δt is laser sheet thickness, A is region of interest (ROI), and M is magnification of the optical system used. Particle dependency of the PIV system has been shown in Fig. E.1.

It might seem after a quick glance that a higher particle density per interrogation window would reduce error, but keeping in mind that particles have a finite dimension, higher density would cause the basic PIV assumption that particles which faithfully follow the fluid flow would get interrupted. Too many particles that produce strong Mie scattering might cause a problem detecting relatively particle-dense regions. Selecting particle density per interrogation window always is a trade-off limited by the induced error as shown in Fig. E.1b. Constant 8–12 particles are kept in all our PIV runs in each window.

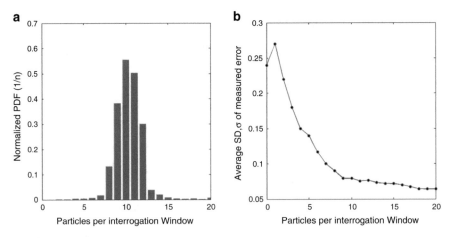

Fig. E.1 Particle dependency of our PIV system has been shown, (**a**) probability density function of particle density per interrogation window and (**b**) average standard deviation of associated errors

Appendix F: DMD Algorithm

Data in the form of a time series, such as images taken in a sequence, can be reconstructed using dynamic mode decomposition (DMD). DMD computes a set of modes that each of which is associated with a fixed oscillation frequency and decay/growth rate. The data can be represented as a snapshot matrix.

$$V_1^N = \{v_1, v_2, v_3, \ldots, v_N\} \tag{F.1}$$

where v_i denotes the ith flow field. Assume a linear mapping A connects the flow field

$$v_{i+1} = Av_i \tag{F.2}$$

Therefore,

$$V_1^N = \{v_1, Av_1, A^2v_1, \ldots, A^{N-1}v_1\} \tag{F.3}$$

Assume there is a specific number N, beyond that the vector v_N can be expressed as the linear combination of the previous vectors

$$v_N = a_1v_1 + a_2v_2 + \ldots + a_{N-1}v_{N-1} = V_1^{N-1}a + r \tag{F.4}$$

Hence,

$$AV_1^{N-1} = V_2^{N-1} = V_1^{N-1}S + re_{N-1}^T \tag{F.5}$$

where r is the vector of residuals that accounts for behaviors that cannot be described completely by A and S is the companion matrix that can be decomposed using eigenvalue.

$$S = T^{-1}\Sigma T \tag{F.6}$$

The eigenvector of S to construct the dynamics modes is $B = V_1^{N-1}T^{-1}$. Therefore, the snapshots based on data set can be decomposed to

$$V_1^{N-1} = BT \tag{F.7}$$

Index

© Springer International Publishing AG, part of Springer Nature 2018
S. Biswas, *Physics of Turbulent Jet Ignition*, Springer Theses,
https://doi.org/10.1007/978-3-319-76243-2

Printed in the United States
By Bookmasters